Leckie × Leckie

Scotland's leading educationa



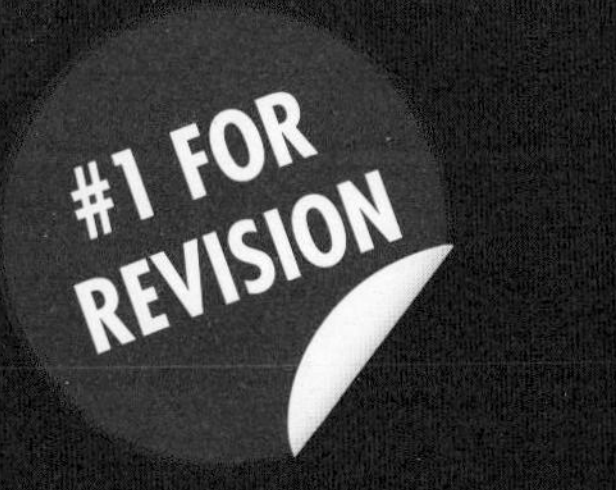

# CfE Higher MATHS SUCCESS GUIDE

n Nisbet

001/20092014

10 9 8 7 6 5 4 3 2 1



ISBN 9780007554362

*Published by*
Leckie & Leckie Ltd
An imprint of HarperCollins*Publishers*
Westerhill Road, Bishopbriggs, Glasgow, G64 2QT
T: 0844 576 8126    F: 0844 576 8131
leckieandleckie@harpercollins.co.uk    www.leckieandleckie.co.uk

Publisher: Peter Dennis
Project Manager: Craig Balfour

*Special thanks to*
Ink Tank (cover design); QBS (layout and illustration); Louise Robb (copy editing); Peter Lindsay (proofreading)

A CIP Catalogue record for this book is available from the British Library.

*Acknowledgements*
Whilst every effort has been made to trace the copyright holders, in cases where this has been unsuccessful, or if any have inadvertently been overlooked, the Publishers would gladly receive any information enabling them to rectify any error or omission at the first opportunity.

Printed and bound by L.E.G.O. S.p.A., Italy

# Contents

## Contents

# This book, your course and your exam

## Using this book

**TOP TIP**

Always revisit questions you failed to solve. Wait a few days then try to solve them again. Use this guide to help!

### About this book

This success guide was written to help you understand and revise the Higher Mathematics course. It will help you pass the three unit tests and the end-of-course exam. The guide follows the unit structure of the course as detailed in the SQA National Course Specifcation Document.

However, you don't become good at mathematics just by reading a guide like this one, although it will give you the knowledge and skills that you need. To become good at mathematics you will need to practice problems – the more you solve problems the better you will become. So use this book as a starting point – come back to it for the essential knowledge and skills that you need to tackle questions and solve problems. If you get stuck there will be a similar question in this guide to help you.

### Top tips

Throughout this guide you will fnd "top tips" about the mathematics you are learning and about exam technique. You should study these carefully

### Quick tests

You will find the "quick tests" after each topic. If you have diffculty with the questions in a test then you need to continue to revise that topic. The answers to the tests are at the back of the guide.

### Sample assessment questions

At the end of each unit you will fnd a selection of typical questions from the unit tests and the end-of-course exam. Detailed solutions to these questions are to be found on the web page: www.leckieandleckie.co.uk/highermathssuccess You should spend time attempting these questions and then compare your solutions carefully with the given solutions.

Use the "quick tests" and sample assessment questions to identify your strengths and weaknesses

## Exam practice

The best preparation for sitting an exam is to try a practice paper under exam conditions at home and then compare your solutions with exemplar solutions. You should obtain Leckie-and-Leckie's Higher Maths Practice Papers for this purpose. They contain typical exam papers with detailed solutions, helpful comments and useful guidance. You will fnd more details on the website: www.leckieandleckie.co.uk

The last few years' actual higher exam papers can be downloaded directly from the SQA website: www.sqa.org.uk/pastpapers Marking schemes for these papers are also available but be warned that they are not nearly as user-friendly as the guidance given in the Leckie-and-Leckie practice paper book. These marking schemes are intended for SQA markers.

**TOP TIP**

When revising for a maths exam **all** your time should be spent doing questions. Use this guide to help you understand how to solve these questions.

## The course structure

The Higher Mathematics course consists of three units:

| | |
|---|---|
| Mathematics: Expressions and Functions (Higher) | Unit 1 of this guide |
| Mathematics: Relationships and Calculus (Higher) | Unit 2 of this guide |
| Mathematics: Applications (Higher) | Unit 3 of this guide |

Success at this course will allow you to progress to the Advanced Higher Mathematics course.

## The assessment structure

To gain a course award you have to pass all three unit tests as well as the end-of-course exam. You should realise that the unit tests will assess only the less demanding parts of the course and that unit test questions are very predictable. So if you get very good at unit test questions this does not necessarily mean you will pass the end-of-course exam.

Your unit tests will be sat on a unit-by-unit basis when your teacher decides you are ready. There are no grades given, only a pass or fail. You will be allowed to use a calculator during all your unit tests. Normally only one attempt at a resit is allowed should you fail a unit test.

The end-of-course exam will consist of two papers as follows:

| | | |
|---|---|---|
| Paper 1 (non-calculator) | 70 minutes | worth 60 marks |
| Paper 2 (calculator allowed) | 90 minutes | worth 70 marks |

Both papers will consist of some short response and some extended response questions.

Depending on how well you performed in this end-of-course-exam (and provided you passed all three unit tests) you will be awarded a grade A,B,C or D

Leckie-and-Leckie produce a set of Higher Mathematics Practice Papers with step-by-step clearly explained solutions which form an excellent resource for exam practice.

## The formulae list

The following formulae list will be available to you during your unit tests and during your end-ofcourse exam:

### FORMULAE LIST

**Circle:**

The equation $x^2 + y^2 + 2gx + 2fy + c = 0$ represents a circle centre $(-g, -f)$ and radius $\sqrt{g^2 + f^2 - c}$.

The equation $(x - a)^2 + (y - b)^2 = r^2$ represents a circle centre $(a, b)$ and radius $r$.

**Scalar Product :** $\mathbf{a.b} = |\mathbf{a}||\mathbf{b}| \cos\theta$, where $\theta$ is the angle between $\mathbf{a}$ and $\mathbf{b}$

or $\mathbf{a.b} = a_1b_1 + a_2b_2 + a_3b_3$ where $\mathbf{a} = \begin{pmatrix} a_1 \\ a_2 \\ a_3 \end{pmatrix}$ and $\mathbf{b} = \begin{pmatrix} b_1 \\ b_2 \\ b_3 \end{pmatrix}$

**Trigonometric formulae:**

$$\sin(A \pm B) = \sin A \cos B \pm \cos A \sin B$$
$$\cos(A \pm B) = \cos A \cos B \mp \sin A \sin B$$
$$\sin 2A = 2\sin A \cos A$$
$$\cos 2A = \cos^2 A - \sin^2 A$$
$$= 2\cos^2 A - 1$$
$$= 1 - 2\sin^2 A$$

**Table of standard derivatives:**

| $f(x)$ | $f'(x)$ |
|---|---|
| $\sin ax$ | $a\cos ax$ |
| $\cos ax$ | $-a\sin ax$ |

**Table of standard integrals:**

| $f(x)$ | $\int f(x)dx$ |
|---|---|
| $\sin ax$ | $-\frac{1}{a}\cos ax + C$ |
| $\cos ax$ | $\frac{1}{a}\sin ax + C$ |

# Some preliminary notation

## Set notation

Sets are collections of things, usually numbers. The **members** or **elements** of a set are **listed** or **described** inside 'curly brackets' **{ }**.

$\in$ means 'is a member of'.

$\notin$ means 'is not a member of'.

The **Empty Set** is the set with no members.

A collection of only some of the members of a given set is called a **subset** of that set.

## Sets of numbers

You will need to recognise the following sets of numbers:

**N** = **{Natural Numbers}** = {1, 2, 3, 4, ...}

**W** = **{Whole Numbers}** = {0, 1, 2, 3, ...}

**Z** = **{Integers}** = $\{\underbrace{\ldots, -3, -2, -1}_{\text{Negative Integers}}, 0, \underbrace{1, 2, 3, \ldots}_{\text{Positive Integers}}\}$

**Q** = **{Rational Numbers}**

These are numbers that can be written as a 'Ratio' of two integers, eg $\frac{2}{3}$, $-4 = \frac{-4}{1}$, $1{\cdot}25 = \frac{5}{4}$

**R** = **{Real Numbers}**

These are numbers that can be represented by **all** the points on the Real Number line, e.g.

$-3$ $-\sqrt{3}$ $0$ $\frac{1}{3}$ $\sqrt{2}$ $2$ $\pi$

**Note:**

**N** is a subset of **W**, **W** is a subset of **Z**, **Z** is a subset of **Q**, **Q** is a subset of **R**.

**Examples**

$2 \in$ **N**. This means '2 is a Natural Number'.

The Even Numbers and the Odd Numbers are subsets of the Natural Numbers.

$1 \notin$ {Primes}. This means '1 is not a Prime Number'.

The set of solutions from **R** of $x^2 = -4$ is the Empty Set. In other words the equation has no Real solutions.

# Functions and their graphs

## What is a function?

A function, $f$, consists of:

1. a **formula**, $f(x)$, which tells you what to do with a given value of $x$.
2. a **domain** which describes the values of $x$ you are allowed to use in the formula.

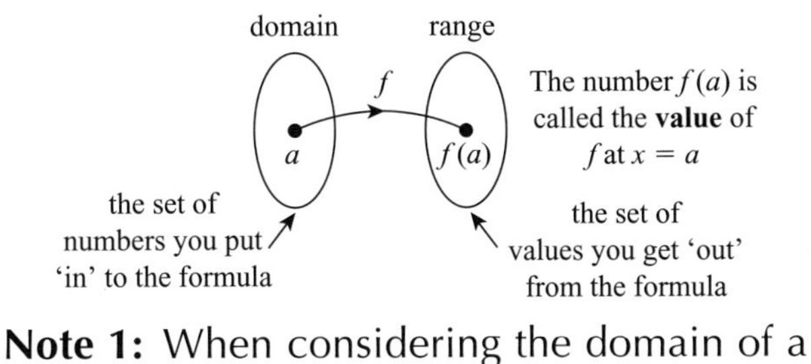


**Note 1:** When considering the domain of a given function **avoid** numbers that will cause:

- Division by zero
- Square-rooting a negative number.

**Note 2:** The **range** is the set of all possible values of the function.

**Examples**

Describe a suitable domain for the functions defined by:

a) $f(x) = \frac{x+1}{x^2+x-6}$ b) $g(x) = \sqrt{x-3}$

**Solution**

a) Avoid $x^2 + x - 6 = 0$
$(x - 2)(x + 3) = 0$
$x = 2$ or $x = -3$
A suitable domain is: all real numbers apart from $-3$ and $2$

b) Avoid $x - 3 < 0$ so avoid $x < 3$
A suitable domain is: all real numbers $x \geq 3$

## Function graphs

A typical **graph** of a function $f$ shows the points $(a, f(a))$ for all values $x = a$ in the domain of $f$.

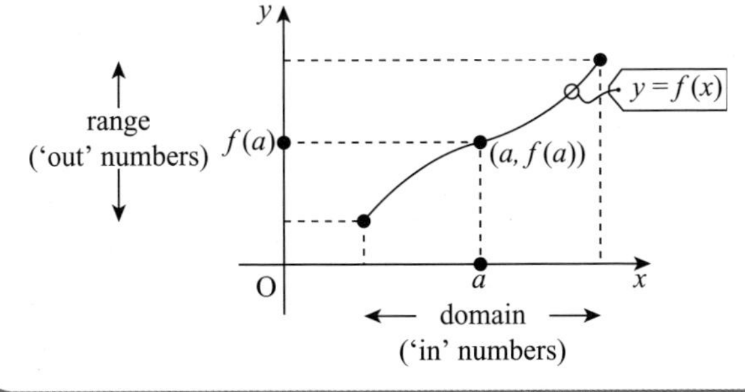


**Example**

The point $(2,k)$ lies on the graph $y = f(x)$ where $f(x) = 3x - 1$

Find the value of $k$.

**Solution:**

$f(2) = 3 \times 2 - 1 = 5$ so $k = 5$ ((2,5) lies on the graph)

**TOP TIP**

For a point on the graph: the value of the $x$-coordinate is put into the formula of the function $f(x)$ to get the value of the $y$-coordinate

# Quick Test 1

1. State a suitable domain for the function $f$. a) $f(x) = \frac{x+1}{x+2}$ b) $f(x) = \sqrt{10-x}$

2. In each case the given point lies on the graph $y = f(x)$. Find the value of $b$.

a) $(3, b)$, $f(x) = 2x + 3$ b) $(-4, b)$ $f(x) = x^2 - x - 1$

# Composite and inverse functions

## Composite functions

Two functions $f$ and $g$ can be combined 'one after the other':

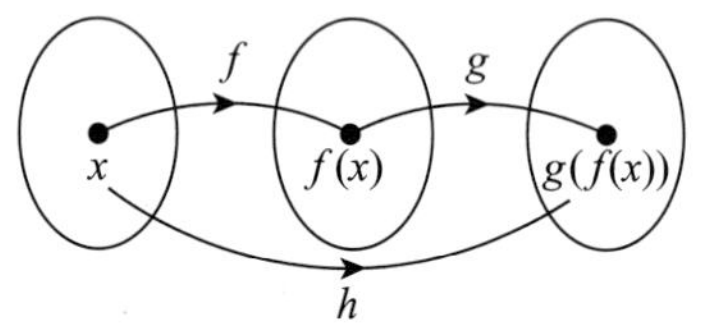


This gives a new function $h$ defined by

$$h(x) = g(f(x))$$

**Example**

If $f(x) = 2x - 1$ and $g(x) = 3x^2$, find $f(g(x))$ and $g(f(x))$ and show $g(f(x)) - 2f(g(x)) = 5 - 12x$

**Solution**

$f(g(x)) = f(3x^2) = 2(3x^2) - 1 = \mathbf{6x^2 - 1}$

$g(f(x)) = g(2x - 1) = 3(2x - 1)^2$
$= \mathbf{12x^2 - 12x + 3}$

So $g(f(x)) - 2f(g(x))$
$= 12x^2 - 12x + 3 - 2(6x^2 - 1)$
$= 12x^2 - 12x + 3 - 12x^2 + 2 = \mathbf{5 - 12x}$

## Inverse functions

A function $f$ can have an inverse $\boldsymbol{f^{-1}}$ which 'undoes' $f$:

$$f^{-1}(f(a)) = a$$

for all values $x = a$ in the domain of $f$.

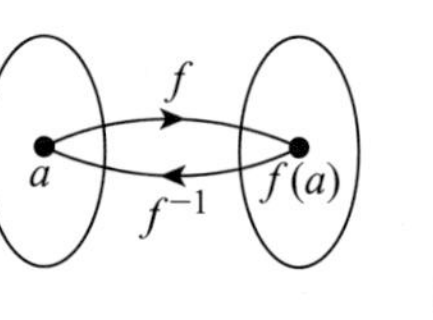


**TOP TIP**

Some simple inverses:
$f(x) = x + k \Rightarrow f^{-1}(x) = x - k$
$f(x) = x - k \Rightarrow f^{-1}(x) = x + k$
$f(x) = kx \Rightarrow f^{-1}(x) = \frac{x}{k}$
$f(x) = \frac{x}{k} \Rightarrow f^{-1}(x) = kx$
where $k$ is a constant.

## Inverse of a linear function

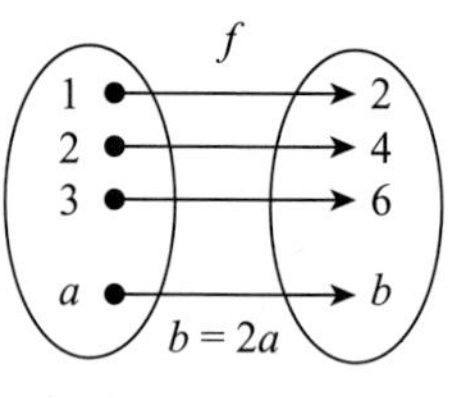


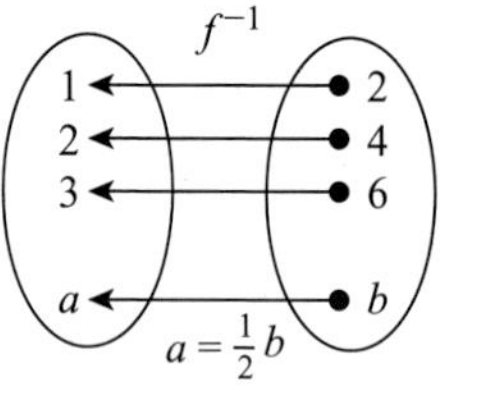


The inverse of the *doubling* function is the *halving* function.

You change the subject of $b = 2a$ from $b$ to $a$ to get $a = \frac{1}{2}b$

So when $f(x) = 2x$ then $f^{-1}(x) = \frac{1}{2}x$

To find the inverse of function $f$:

**Step 1** Replace $x$ by $a$ in the formula for $f$. In other words find $f(a)$

**Step 2** Let $b = f(a)$

**Example**

In each case find the inverse function $f^{-1}$

a) $f(x) = \frac{2x}{3}$ b) $f(x) = 5 - 3x$

**Solution**

a) $f(a) = \frac{2a}{3}$ so $b = \frac{2a}{3} \Rightarrow 3b = 2a$

$\Rightarrow \frac{3b}{2} = a$ so $f^{-1}(x) = \frac{3x}{2}$

b) $f(a) = 5 - 3a$ so $b = 5 - 3a$

$\Rightarrow 3a = 5 - b$

$\Rightarrow a = \frac{5-b}{3}$ so $f^{-1}(x) = \frac{5-x}{3}$

**Step 3** Change the subject of the formula in Step 2 from $b$ to $a$

**Step 4** Change $b$ to $x$ to get the inverse formula

## Graphs and inverse functions

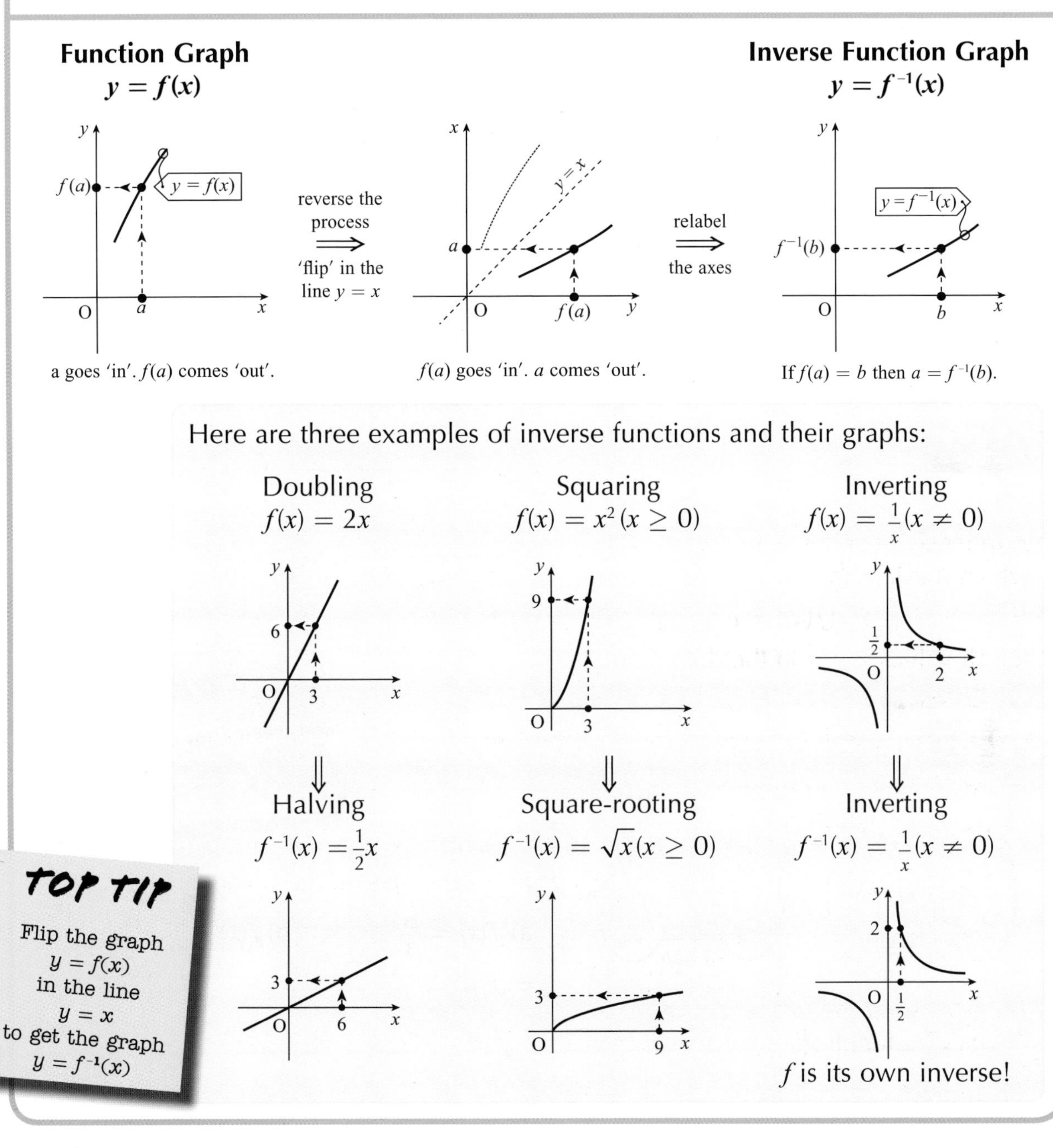


**TOP TIP**

Flip the graph $y = f(x)$ in the line $y = x$ to get the graph $y = f^{-1}(x)$

# Quick Test 2

**1.** $f(x) = 3x - 1$ and $g(x) = x^2$, find:

a) $g(f(x))$ b) $f(g(x))$ c) $f(f(x))$ d) $g(g(x))$

**2.** In each case find the inverse function $f^{-1}$

a) $f(x) = \frac{1}{2}x + 1$ b) $f(x) = 7 - x$ c) $f(x) = \sqrt{x - 1}$

# Quadratic functions

## Completing the square

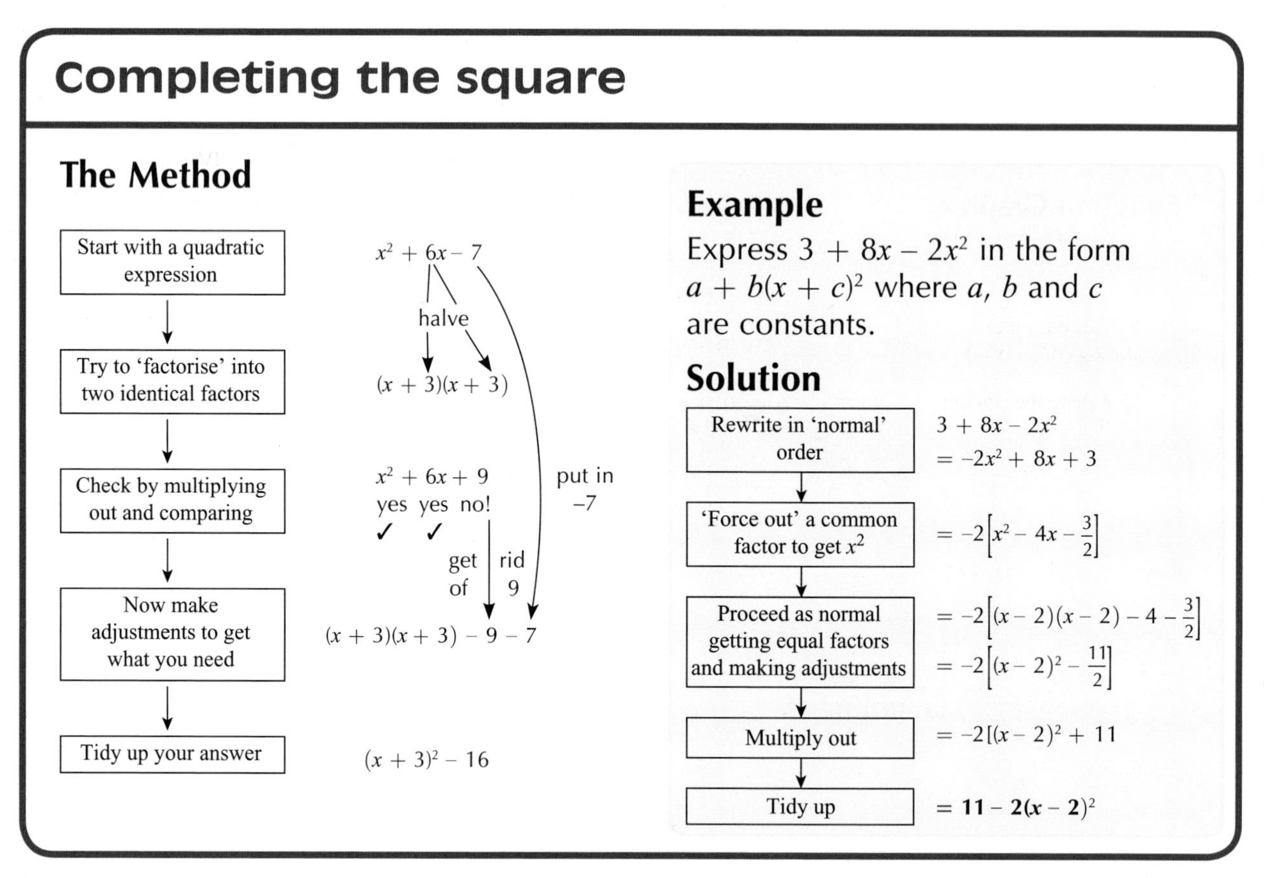


## Graphs related to $y = x^2$

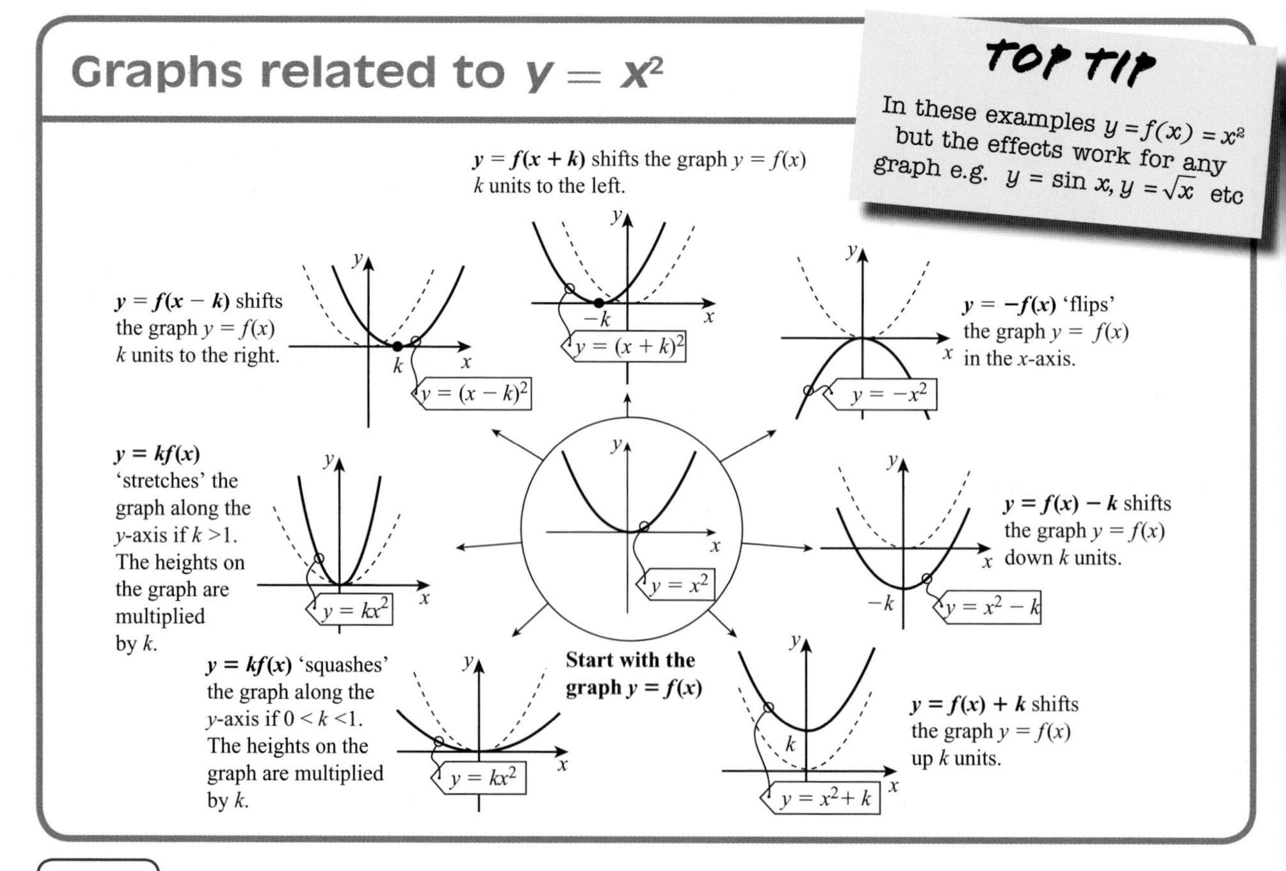

## Sketching quadratic graphs

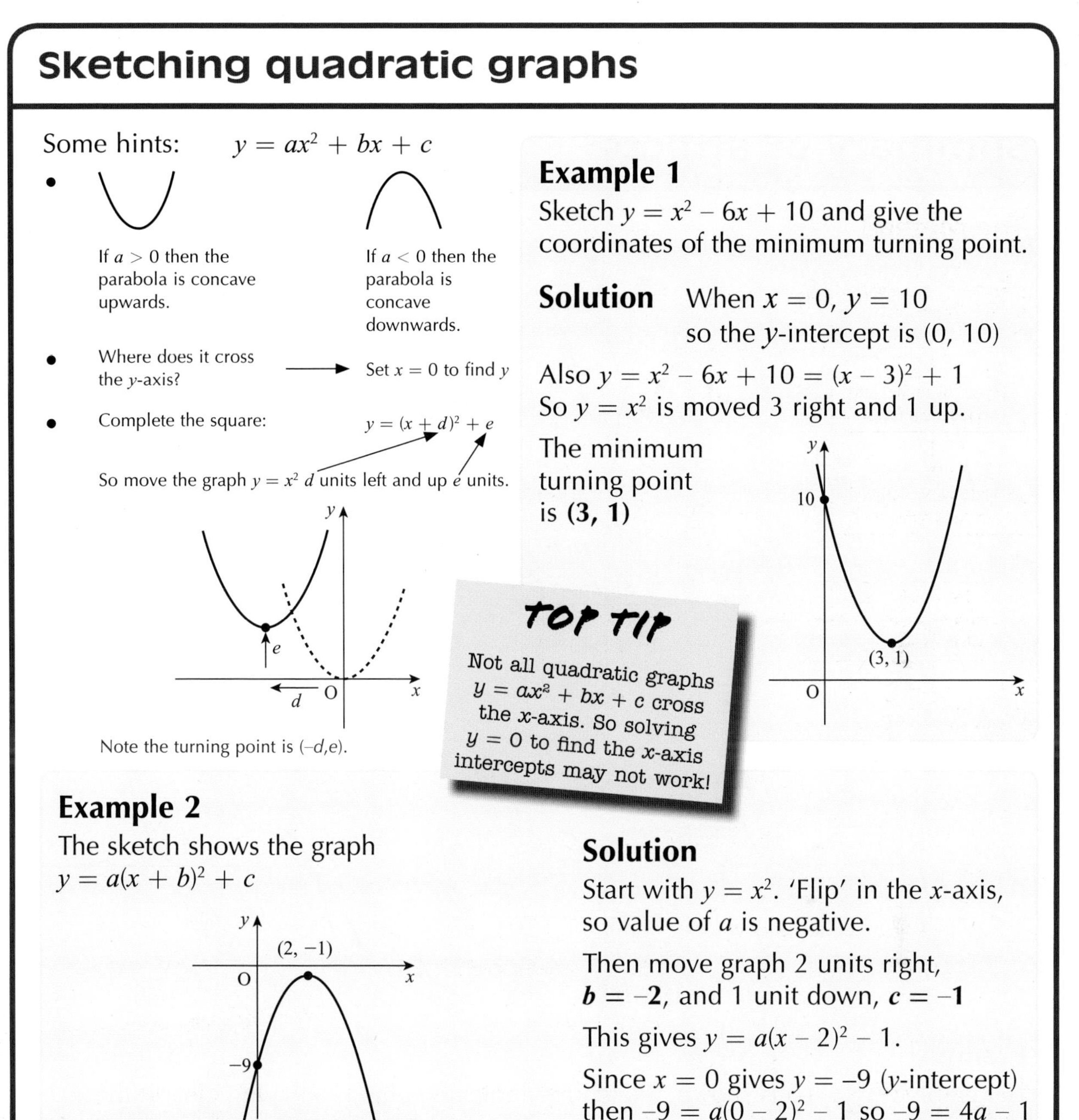

Some hints: $y = ax^2 + bx + c$

- If $a > 0$ then the parabola is concave upwards. If $a < 0$ then the parabola is concave downwards.
- Where does it cross the $y$-axis? → Set $x = 0$ to find $y$
- Complete the square: $y = (x + d)^2 + e$

So move the graph $y = x^2$ $d$ units left and up $e$ units.



Note the turning point is $(-d, e)$.

**Example 1**

Sketch $y = x^2 - 6x + 10$ and give the coordinates of the minimum turning point.

**Solution** When $x = 0$, $y = 10$ so the $y$-intercept is $(0, 10)$

Also $y = x^2 - 6x + 10 = (x - 3)^2 + 1$

So $y = x^2$ is moved 3 right and 1 up.

The minimum turning point is **(3, 1)**



**TOP TIP**

Not all quadratic graphs $y = ax^2 + bx + c$ cross the $x$-axis. So solving $y = 0$ to find the $x$-axis intercepts may not work!

**Example 2**

The sketch shows the graph $y = a(x + b)^2 + c$



Find the values of $a$, $b$ and $c$.

**Solution**

Start with $y = x^2$. 'Flip' in the $x$-axis, so value of $a$ is negative.

Then move graph 2 units right, $\boldsymbol{b = -2}$, and 1 unit down, $\boldsymbol{c = -1}$

This gives $y = a(x - 2)^2 - 1$.

Since $x = 0$ gives $y = -9$ ($y$-intercept) then $-9 = a(0 - 2)^2 - 1$ so $-9 = 4a - 1$ so $4a = -8$ giving $\boldsymbol{a = -2}$

# Quick Test 3

1. Express in the form $a(x + b)^2 + c$ and state the values of $a$, $b$ and $c$
   a) $2x^2 - 4x + 6$
   b) $3x^2 - 6x + 2$
2. The graph shows $y = k(x + m)^2 + n$ find the values of $k$, $m$ and $n$

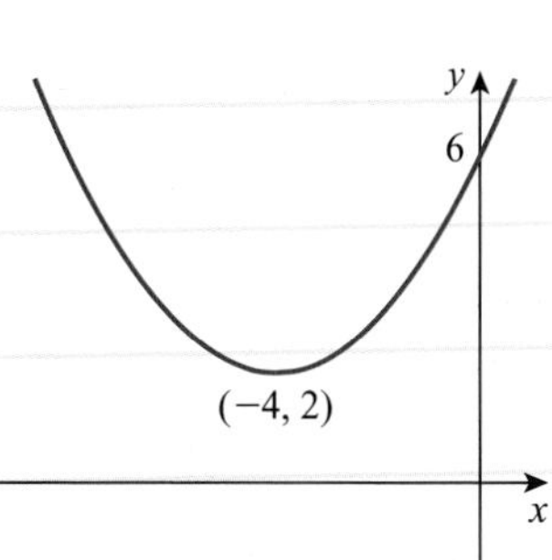

# Related graphs

**TOP TIP**

These effects on the graph $y = f(x)$ are called transformations. The left/right up/down effects are called translations. The flips are called reflections

## Summary of effects

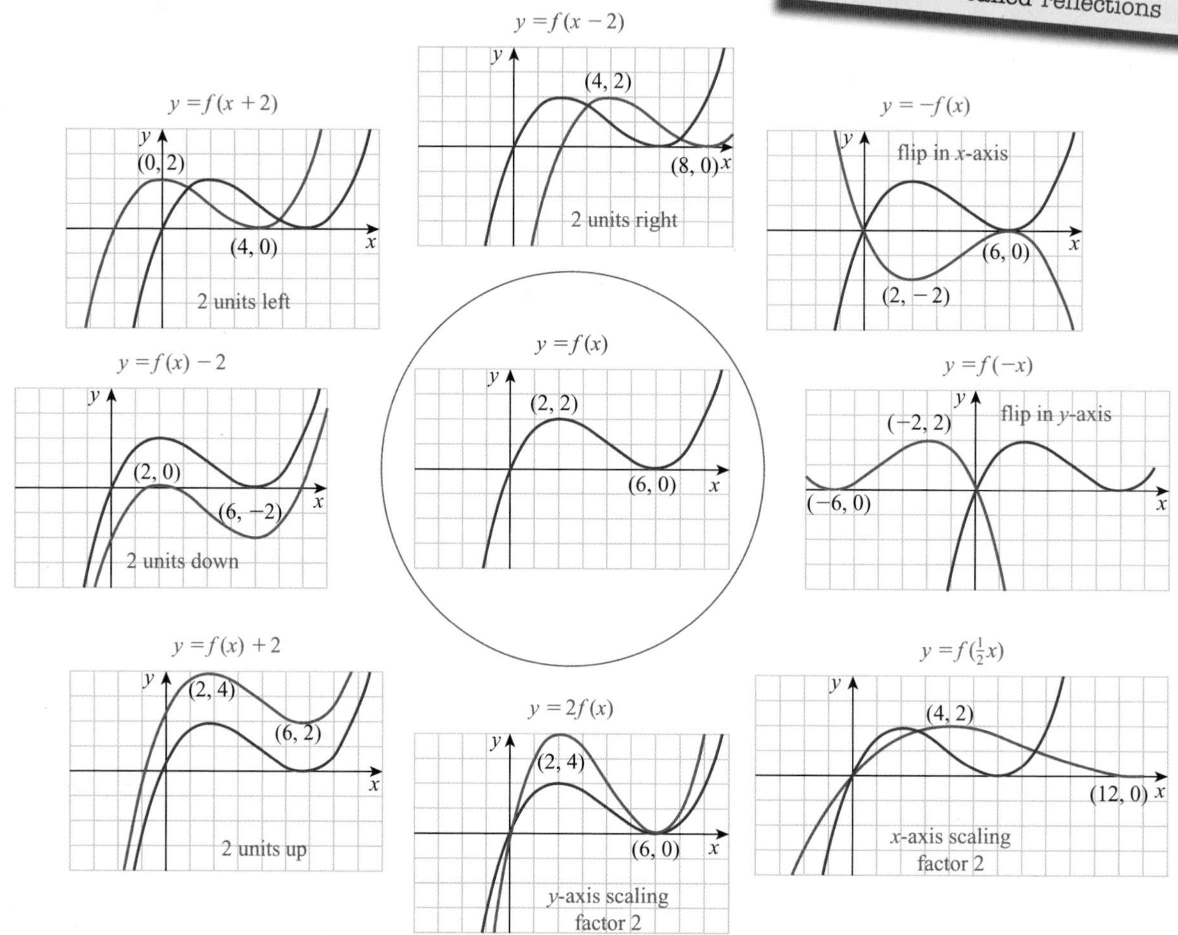


### Example

The graph of $y = f(x)$ is shown. Sketch the indicated graphs showing clearly the images of the named points.

a) $y = f(x + 2)$

b) $y = f(x - 1)$

c) $y = -f(x)$

d) $y = f(x) - 1$

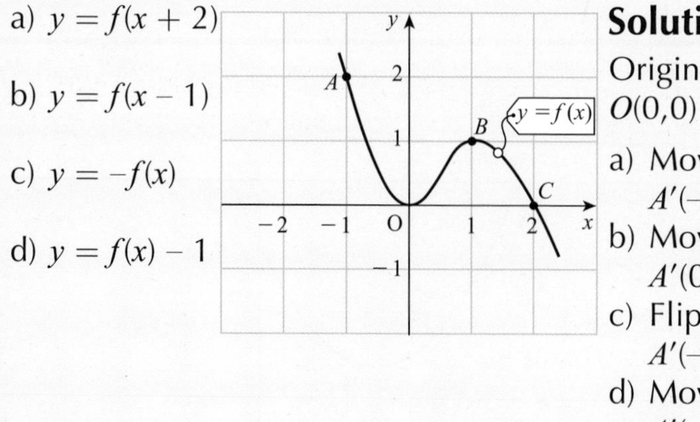


### Solution

Original points: $A(-1,2)$, $B(1,1)$, $C(2,0)$ and $O(0,0)$

a) Move graph 2 units left
$A'(-3, 2)$, $B'(-1, 1)$, $C'(0, 0)$ and $O'(-2, 0)$

b) Move graph 1 unit right
$A'(0, 2)$, $B'(2, 1)$, $C'(3, 0)$ and $O'(1, 0)$

c) Flip graph in $x$-axis
$A'(-1, -2)$, $B'(1, -1)$, $C'(2, 0)$ and $O'(0, 0)$

d) Move graph down 1 unit
$A'(-1, 1)$, $B'(1, 0)$, $C'(2, -1)$ and $O'(0, -1)$

## Combining effects

When several effects are combined you will need to split up this combined effect into individual effects. Some typical splits you might encounter are explained below. Usually you are given the graph $y = f(x)$ and your aim is to draw a related graph. You will also know a few points on the graph $y = f(x)$ and you should be able to find where these points are on the related graph.

Note: $A$ and $B$ are on $y = f(x)$ and $A'$ and $B'$ are on the related graph.

| | | | |
|---|---|---|---|
| Aim: | $y = 3f(x - 5)$ | | |
| Start: | $y = f(x)$ | $A(4, 2)$ | $B(-4, -1)$ |
| 5 units right: | $y = f(x - 5)$ | $(9, 2)$ | $(1, -1)$ |
| $y$-axis scaling (factor 3): | $y = 3f(x - 5)$ | $A'(9, 6)$ | $B'(1, -3)$ |
| Aim: | $y = -\frac{1}{2}f(x + 1) - 3$ | | |
| Start: | $y = f(x)$ | $A(3, 6)$ | $B(0, -1)$ |
| 1 unit left: | $y = f(x + 1)$ | $(2, 6)$ | $(-1, -1)$ |
| $y$-axis scaling (factor $\frac{1}{2}$): | $y = \frac{1}{2}f(x + 1)$ | $(2, 3)$ | $(-1, -\frac{1}{2})$ |
| flip in $x$-axis: | $y = -\frac{1}{2}f(x + 1)$ | $(2, -3)$ | $(-1, \frac{1}{2})$ |
| 3 units down: | $y = -\frac{1}{2}f(x + 1) - 3$ | $A'(2, -6)$ | $B'(-1, -\frac{5}{2})$ |

**TOP TIP**

If you know the coordinates of a point on the graph $y = f(x)$ then a corresponding point should appear on your sketch of the related graph.

**Example**

The graph of $y = f(x)$ is shown:

Sketch the graph $y = 2 - f(x)$

**Solution**

Split into two steps:

**Step 1** graph of $y = -f(x)$

'flip' in $x$-axis

**Step 2** graph of $y = 2 - f(x)$

shift up 2 units

Note: $2 - f(x)$ can be rewritten as $-f(x) + 2$ resulting in moving the graph $y = -f(x)$ up 2 units

# Quick Test 4

The graph of $y = f(x)$ is shown. Sketch the indicated graphs showing clearly the images of the named points.

1. $y = -f(x) + 2$
2. $y = f(x + 1) - 1$

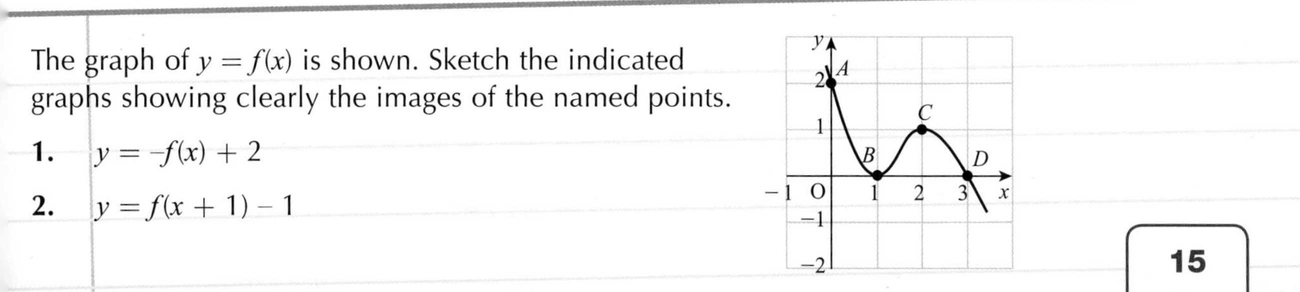

# Exponential functions

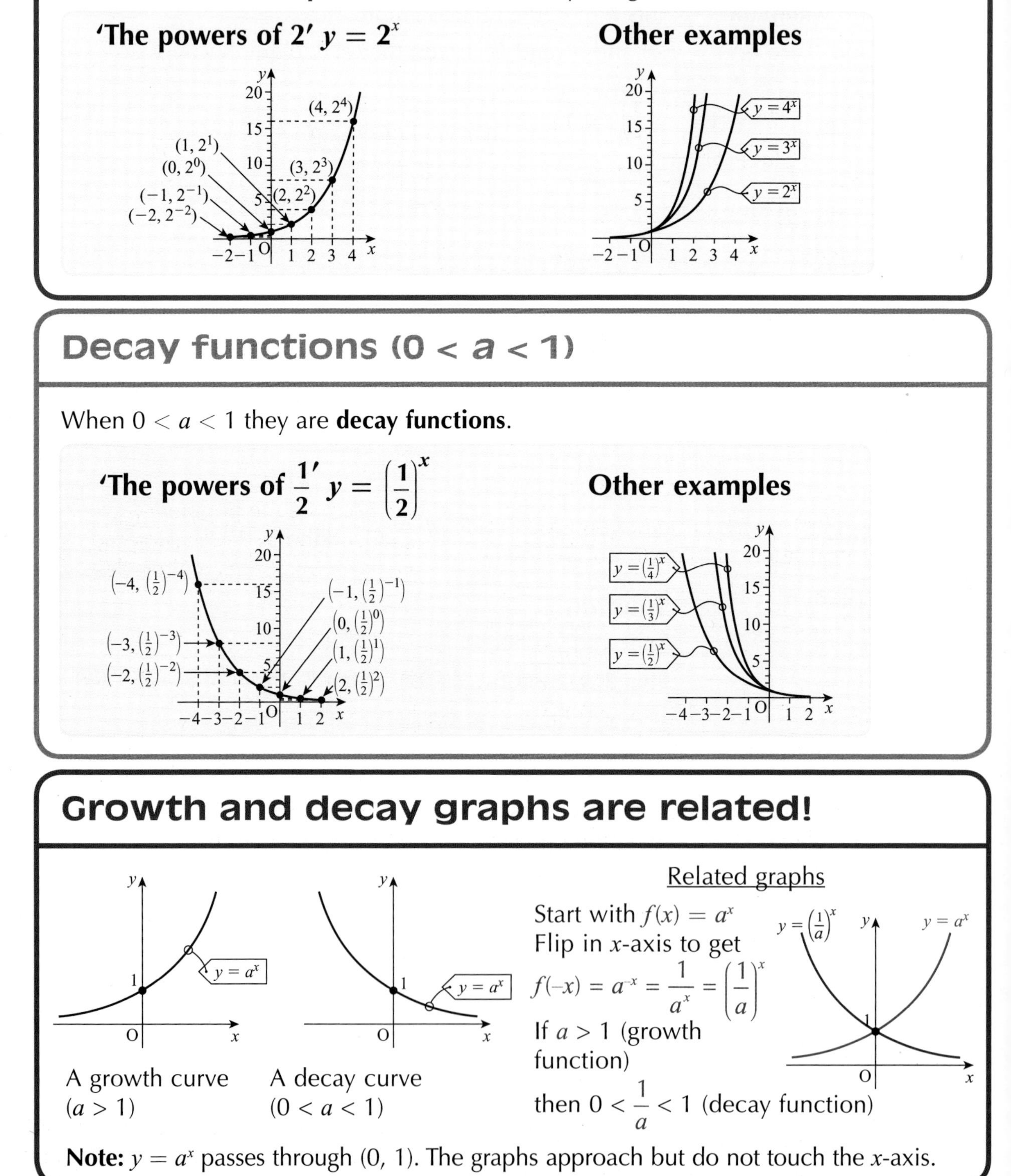


## Growth functions ($a > 1$)

The functions defined by $f(x) = a^x$ are called **exponential functions**. The number $a$ is the **base** and $x$ is the **exponent**. When $a > 1$ they are **growth functions**.

**'The powers of 2' $y = 2^x$**

**Other examples**

## Decay functions ($0 < a < 1$)

When $0 < a < 1$ they are **decay functions**.

**'The powers of $\frac{1}{2}$' $y = \left(\frac{1}{2}\right)^x$**

**Other examples**

## Growth and decay graphs are related!

A growth curve ($a > 1$)

A decay curve ($0 < a < 1$)

Related graphs

Start with $f(x) = a^x$

Flip in $x$-axis to get

$$f(-x) = a^{-x} = \frac{1}{a^x} = \left(\frac{1}{a}\right)^x$$

If $a > 1$ (growth function)

then $0 < \frac{1}{a} < 1$ (decay function)

**Note:** $y = a^x$ passes through (0, 1). The graphs approach but do not touch the $x$-axis.

## Reminder: the rules of indices

| Rule | Examples |
|---|---|
| $x^m \times x^n = x^{m+n}$ | $a^2 \times a^3 = a^{2+3} = a^5$ |
| $\frac{x^m}{x^n} = x^{m-n}$ | $\frac{c^7}{c^3} = c^{7-3} = c^4$ |
| $(x^m)^n = x^{mn}$ | $(y^3)^4 = y^{3 \times 4} = y^{12}$ |
| $x^0 =$ | $2^0 = 1$ $\left(\frac{1}{2}\right)^0 = 1$ $(a+b)^0 = 1$ |
| $x^{-n} = \frac{1}{x^n}$ | $a^{-1} = \frac{1}{a^1} = \frac{1}{a}$ $a^{-3} = \frac{1}{a^3}$ |
| $x^{\frac{m}{n}} = (\sqrt[n]{x})^m$ | $a^{\frac{3}{2}}$ (power 3, square root) $= (\sqrt{a})^3$ $a^{\frac{2}{3}}$ (power 2, cube root) $= (\sqrt[3]{a})^2$ |

**Example**

$f(x) = a^x$ and $g(x) = \left(\frac{1}{a}\right)^x$

a) If $a = 2^{-1}$ find formulae for $f(x)$ and $g(x)$ in simplest form with positive indices.

b) Hence evaluate $f(-\frac{1}{2}) + g(\frac{3}{2})$ for $a = \frac{1}{2}$

**Solution**

a) $f(x) = (2^{-1})^x = 2^{-1 \times x} = 2^{-x} = \frac{1}{2^x}$

$g(x) = \left(\frac{1}{2^{-1}}\right)^x = (2^1)^x = 2^{1 \times x} = 2^x$

b) So $f\left(-\frac{1}{2}\right) + g\left(\frac{3}{2}\right) = \frac{1}{2^{-\frac{1}{2}}} + 2^{\frac{3}{2}} = 2^{\frac{1}{2}} + 2^{\frac{3}{2}}$

$= \sqrt{2} + (\sqrt{2})^3 = \sqrt{2} + \sqrt{2} \times \sqrt{2} \times \sqrt{2}$

$= \sqrt{2} + 2\sqrt{2} = 3\sqrt{2}$

## Exponential models

Scotland in 2011 had a population growth rate of 0·6% per year and a population of 5·3 million.

Population in 2012: $5{\cdot}3 \times 1{\cdot}006$
Population in 2013: $5{\cdot}3 \times 1{\cdot}006 \times 1{\cdot}006$
$= 5{\cdot}3 \times 1{\cdot}006^2$
Population in 2014: $5{\cdot}3 \times 1{\cdot}006^2 \times 1{\cdot}006$
$= 5{\cdot}3 \times 1{\cdot}006^3$

So the population after $x$ years is given by $5{\cdot}3 \times 1{\cdot}006^x$

This is an exponential function which describes mathematically how the population behaves. It is called a mathematical model.

**Example**

Estimate the population of Scotland in 2020 using the same mathematical model as opposite.

**Solution**

Use $x = 9$ as 2020 is 9 years after 2011

Population $= 5{\cdot}3 \times 1{\cdot}006^9$
$= 5{\cdot}59316...$

Estimate is 5·6 million (to 1 decimal place)

# Quick Test 5

**1.** a) For this graph write down an inequality that $a$ satisfies.

b) The point (–1, 3) lies on the curve. Calculate the value of the base $a$.



**2.** The price £$P_m$ of a house in Edinburgh after $m$ months is given by the exponential model:

$$P_m = P_0 (1{\cdot}0035)^m$$

where $P_0$ is the price at the start of 2014. If a house in Edinburgh is worth £280000 at the start of 2014, estimate its value by mid-2016.

# Logarithmic functions

## The inverse of an exponential function

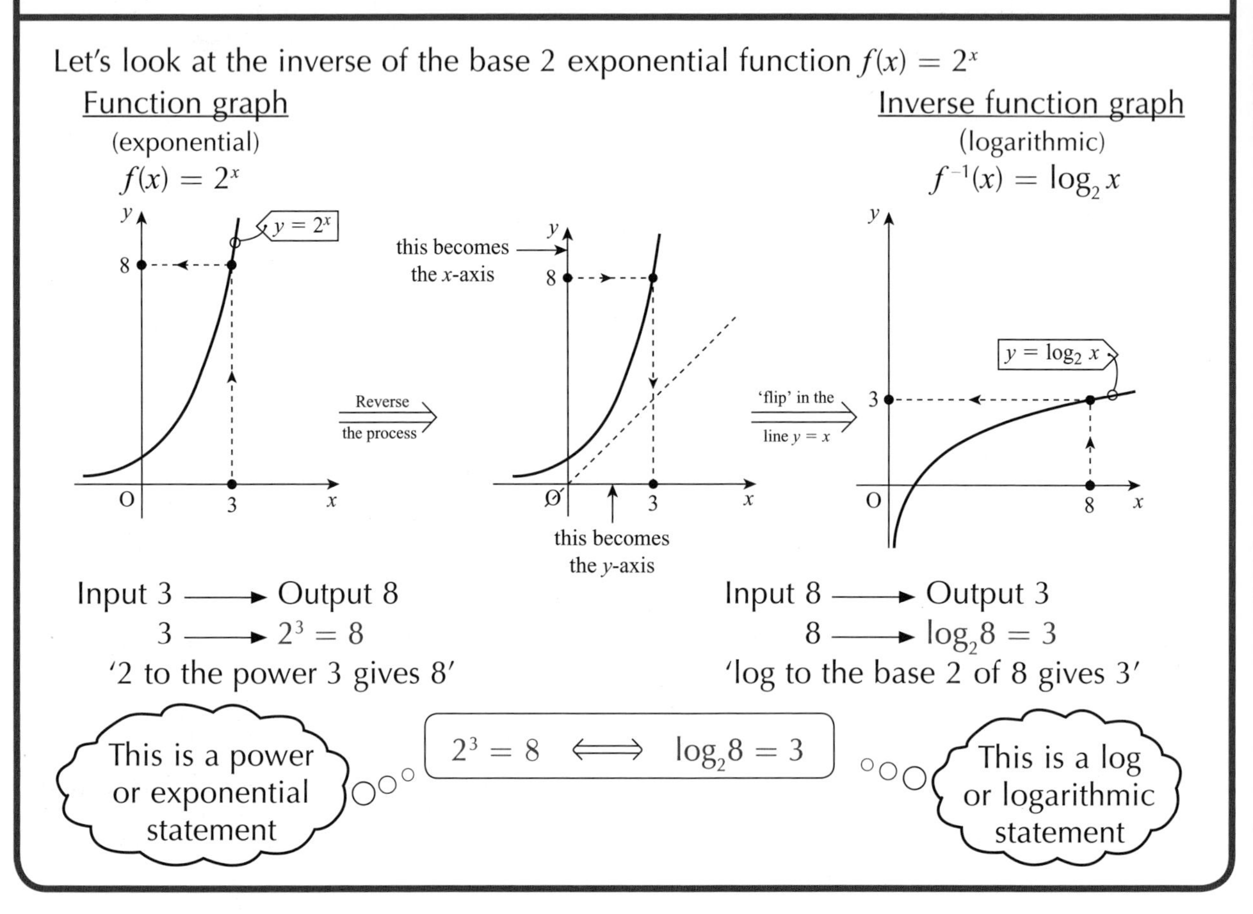


## The logarithmic graph

Here is a typical log graph:

$y$, $y = \log_a x$, $n$, $(m, n)$, O, 1, $m$, $x$

Notes:

1. $(m, n)$ lies on the graph so
   $$n = \log_a m$$
   this 'log statement' can change to
   $$a^n = m$$
   which is a 'power statement'
2. For the graph shown the base $a > 1$
3. Simple log graphs like the one shown pass through the point (1, 0). This is because: (1, 0) on the graph gives $0 = \log_a 1 \Leftrightarrow a^0 = 1$
4. The graph lies entirely to the right of the $y$-axis. $\log_a x$ is not defined for $x \leq 0$. This is because the value of any power of $a$ is always positive.

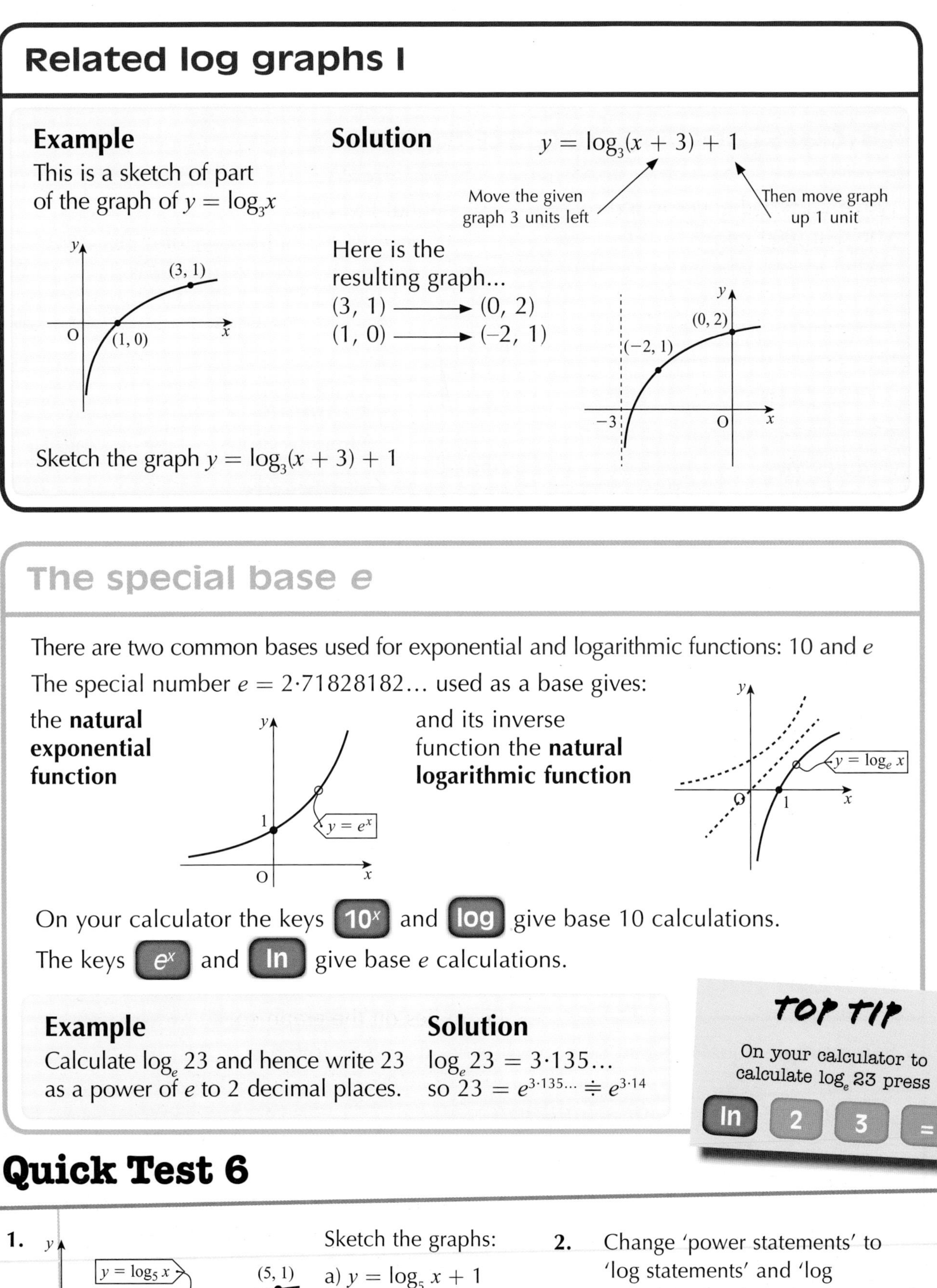

## Related log graphs I

**Example**

This is a sketch of part of the graph of $y = \log_3 x$

Sketch the graph $y = \log_3(x + 3) + 1$

**Solution**

$y = \log_3(x + 3) + 1$

Move the given graph 3 units left

Then move graph up 1 unit

Here is the resulting graph...

$(3, 1) \longrightarrow (0, 2)$

$(1, 0) \longrightarrow (-2, 1)$

## The special base *e*

There are two common bases used for exponential and logarithmic functions: 10 and $e$

The special number $e = 2{\cdot}71828182...$ used as a base gives:

the **natural exponential function** and its inverse function the **natural logarithmic function**

On your calculator the keys $10^x$ and log give base 10 calculations.

The keys $e^x$ and ln give base $e$ calculations.

**Example**

Calculate $\log_e 23$ and hence write 23 as a power of $e$ to 2 decimal places.

**Solution**

$\log_e 23 = 3{\cdot}135...$

so $23 = e^{3{\cdot}135...} \doteqdot e^{3{\cdot}14}$

**TOP TIP**

On your calculator to calculate $\log_e 23$ press ln 2 3 =

# Quick Test 6

1. Sketch the graphs:

a) $y = \log_5 x + 1$

b) $y = \log_5(x + 1)$

c) $y = -\log_5 x$

d) $y = \log_5(x - 2) - 1$

2. Change 'power statements' to 'log statements' and 'log statements' to 'power statements':

a) $92 = 10^x$ b) $4 = e^y$

c) $4 = \log_{10} t$ d) $\log_e A = B$

# Working with logs

## Understanding 'log statements'

Since $2^3 = 8$ changes to $\log_2 8 = 3$ you can read $\log_2 8 = 3$ as:
'What power of 2 gives 8? Answer:3'

For example:

$\log_3 81$: 'What power of 3 gives 81?'
Since $3^4 = 81$ the answer is 4
So $\log_3 81 = 4$

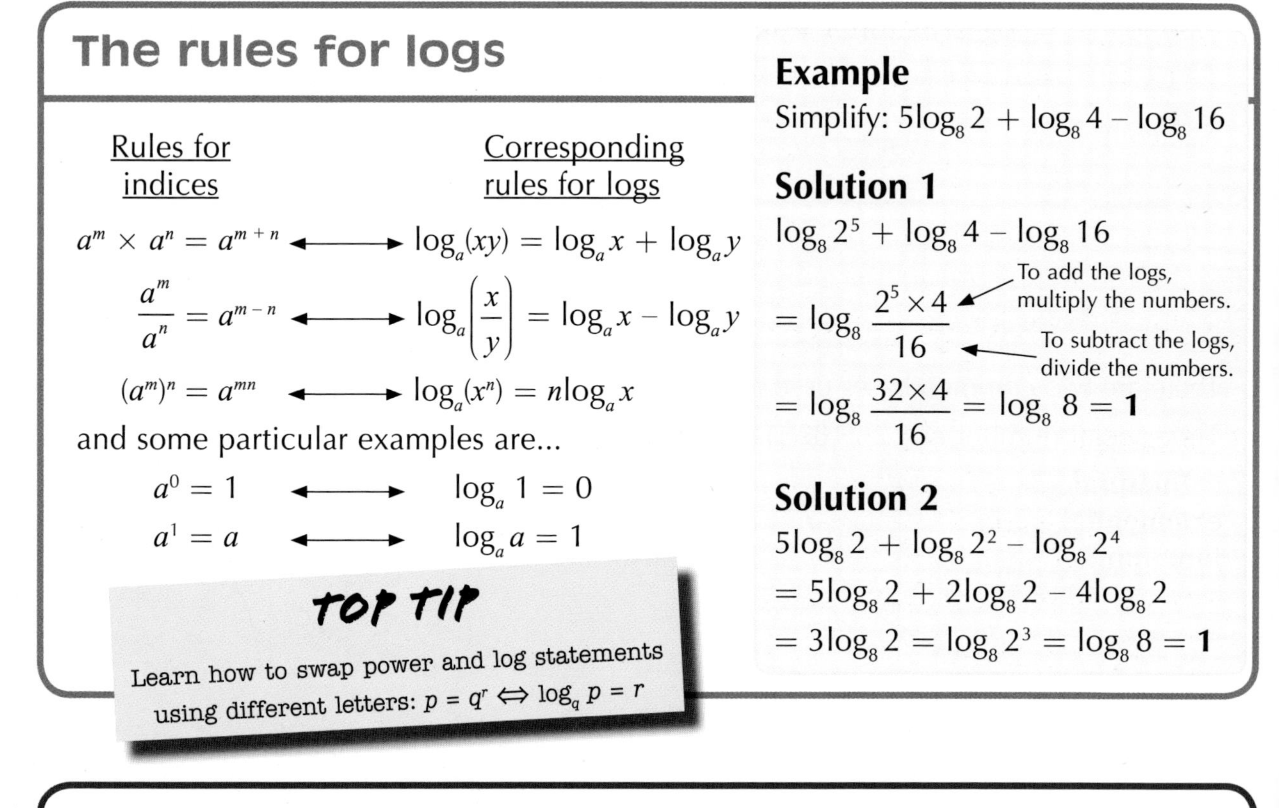

## The rules for logs

| Rules for indices | | Corresponding rules for logs |
|---|---|---|
| $a^m \times a^n = a^{m+n}$ | ⟷ | $\log_a(xy) = \log_a x + \log_a y$ |
| $\dfrac{a^m}{a^n} = a^{m-n}$ | ⟷ | $\log_a\left(\dfrac{x}{y}\right) = \log_a x - \log_a y$ |
| $(a^m)^n = a^{mn}$ | ⟷ | $\log_a(x^n) = n\log_a x$ |

and some particular examples are...

| | | |
|---|---|---|
| $a^0 = 1$ | ⟷ | $\log_a 1 = 0$ |
| $a^1 = a$ | ⟷ | $\log_a a = 1$ |

**TOP TIP**

Learn how to swap power and log statements using different letters: $p = q^r \Leftrightarrow \log_q p = r$

**Example**

Simplify: $5\log_8 2 + \log_8 4 - \log_8 16$

**Solution 1**

$\log_8 2^5 + \log_8 4 - \log_8 16$

$= \log_8 \dfrac{2^5 \times 4}{16}$

To add the logs, multiply the numbers.

To subtract the logs, divide the numbers.

$= \log_8 \dfrac{32 \times 4}{16} = \log_8 8 = \mathbf{1}$

**Solution 2**

$5\log_8 2 + \log_8 2^2 - \log_8 2^4$

$= 5\log_8 2 + 2\log_8 2 - 4\log_8 2$

$= 3\log_8 2 = \log_8 2^3 = \log_8 8 = \mathbf{1}$

## Simple log equations

Some simple equations can be solved using: $b^c = a \Leftrightarrow \log_b a = c$

**Example**

a) $\log_e x = 5$

b) $\log_{10} x = 2{\cdot}9$

c) $e^x = 4{\cdot}5$

d) $10^x = 2$

**Solution**

a) $\log_e x = 5 \Rightarrow x = e^5 \doteqdot 148{\cdot}4$ (using $e^x$)

b) $\log_{10} x = 2{\cdot}9 \Rightarrow x = 10^{2\cdot9} \doteqdot 794{\cdot}3$ (using $10^x$)

c) $e^x = 4{\cdot}5 \Rightarrow x = \ln 4{\cdot}5 \doteqdot 1{\cdot}50$ (using ln)

d) $10^x = 2 \Rightarrow x = \log_{10} 2 \doteqdot 0{\cdot}301$ (using log)

## Taking the logs of both sides

For some problems a useful technique is to 'take the logs' of both sides of an equation.

**Example 1**

Evaluate $\log_3 2$

**Solution**

Let $x = \log_3 2$

Then $3^x = 2$

Now take logs of both sides (base $e$ or 10)

So $\log_e 3^x = \log_e 2 \Rightarrow x \ln 3 = \ln 2$

$\Rightarrow x = \frac{\ln 2}{\ln 3} \doteqdot 0{\cdot}631$ $\left(\text{check } \frac{\log_{10} 2}{\log_{10} 3} \text{ gives the same}\right)$

**Example 2**

Solve $5^x = 4$

**Solution**

Take logs of both sides (either base $e$ **or** base 10)

so $\log_{10}(5^x) = \log_{10} 4$

$\Rightarrow x \log_{10} 5 = \log_{10} 4$

$\Rightarrow x = \frac{\log_{10} 4}{\log_{10} 5} \doteqdot 0{\cdot}861$

## Related log graphs II

Remind yourself how the graphs $y = af(x)$ and $y = f(x + b)$ are obtained from the graph $y = f(x)$

**Example**

The diagram shows the graph $y = a\log_2(x + b)$. Find the values of $a$ and $b$.

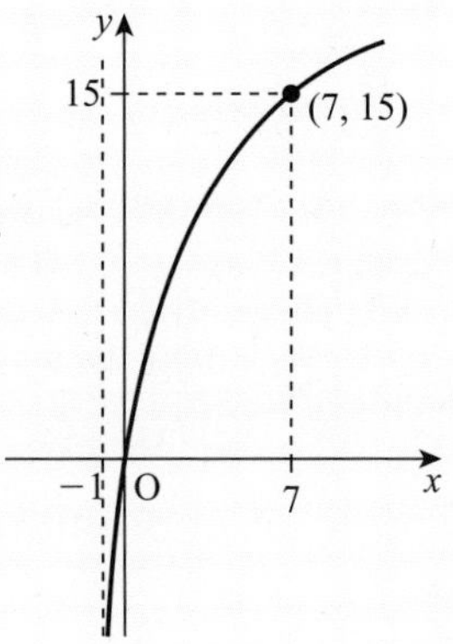

**Solution**

The graph $y = \log_a x$ passes through $(1, 0)$

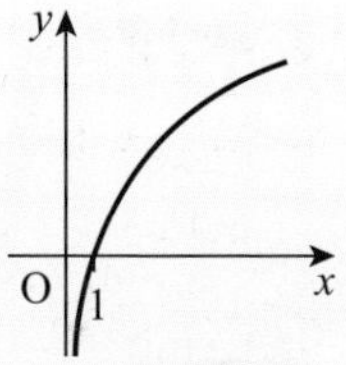

The given graph passes through $(0, 0)$ and so $b = 1$ (the graph above moves 1 unit left) giving $y = a\log_2(x + 1)$

$a$ is a $y$-axis scaling. Use $(7, 15)$ to find $a$

$x = 7$, $y = 15$ satisfy the equation

so $15 = a\log_2(7 + 1)$

$\Rightarrow 15 = a\log_2 8 = a \times 3$

$\Rightarrow 15 = 3a \Rightarrow a = 5$

# Quick Test 7

1. Evaluate $\log_3 27 + \log_2 \frac{1}{2}$
2. Simplify $\log_a x^2 - \log_a x + \log_a 1$
3. Solve a) $e^x = 3{\cdot}2$ b) $7^x = 5$ to 2 decimal places.
4. The diagram shows the graph of $y = f(x)$ where $f(x) = a\log_2(x - b)$

   Find the values of $a$ and $b$.

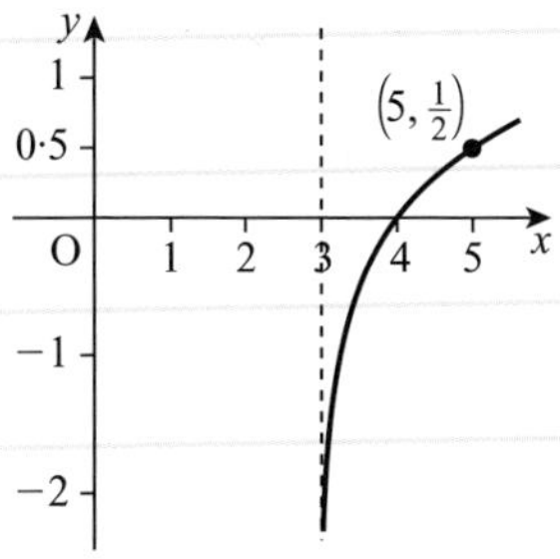

# Problem solving with logs

## Interpreting and solving problems

**Example** The amount $A$ grams of a radioactive substance after $t$ units of time is given by $A = A_0e^{-kt}$ where $A_0$ is the initial amount of the substance and $k$ is a constant.

Calculate the half-life of neptunium if it takes 3 hours for 50 grams to reduce to 45·5 grams.

**Solution** What variable values do you know?

You know $t = 3$, $A_0 = 50$ and $A = 45{\cdot}5$ from which you can find the constant $k$:

$$45{\cdot}5 = 50e^{-3k} \implies \frac{45{\cdot}5}{50} = e^{-3k} \implies \log_e \frac{45{\cdot}5}{50} = -3k$$

rearranging gives $k = \dfrac{\log_e \frac{45\cdot5}{50}}{-3} = 0{\cdot}0314...$

The formula is now: $A = A_0e^{-0.0314... \times t}$

What does half-life mean?

This is the time it takes for the initial amount of the substance to reduce to half that amount.

You have to find $t$ for $A_0$ to reduce to $0{\cdot}5A_0$:

$$0{\cdot}5A_0 = A_0e^{-0{\cdot}0314... \times t} \implies 0{\cdot}5 = e^{-0{\cdot}0314... \times t}$$

Change this to a log statement:

$$\log_e 0{\cdot}5 = -0{\cdot}0314... \times t \implies t = \frac{\log_e 0{\cdot}5}{-0{\cdot}0314...} = 22{\cdot}048...$$

So the required half-life of this sample of neptunium is approximately 22 hours.

**TOP TIP**

Never use rounded values in your calculations. Only round at the end of the calculation.

## Determining data set relationships

Often scientists look at experimental data to find the relationship between two variables. But the rule to calculate the values of $y$ from the values of $x$ may not be obvious even after the values are graphed.

Logs can be used to reveal certain types of relationships:

**$y = ax^b$**

Taking the log of both sides gives:

$$\log y = \log(ax^b) = \log a + \log x^b$$

so $\log y = b \log x + \log a$

compare $Y = mX + c$

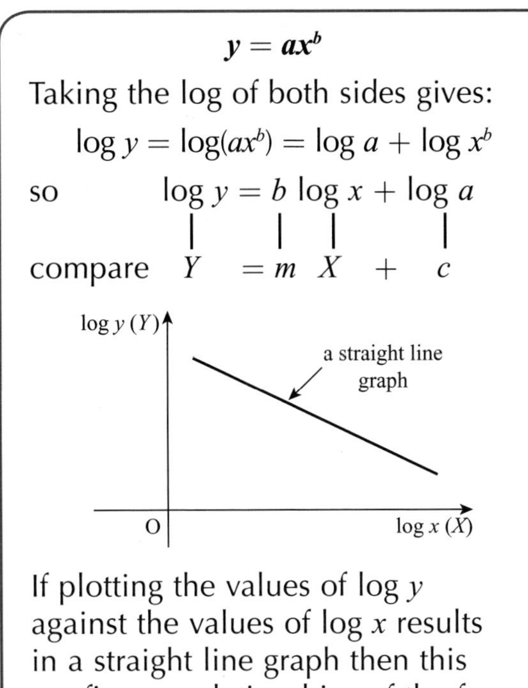


If plotting the values of log $y$ against the values of log $x$ results in a straight line graph then this confirms a relationships of the form

$$y = ax^b$$

for suitable values of constants $a$ and $b$.

**$y = ab^x$**

Taking the log of both sides gives:

$$\log y = \log(ab^x) = \log a + \log b^x$$

so $\log y = (\log b)\, x + \log a$

compare $Y = m x + c$

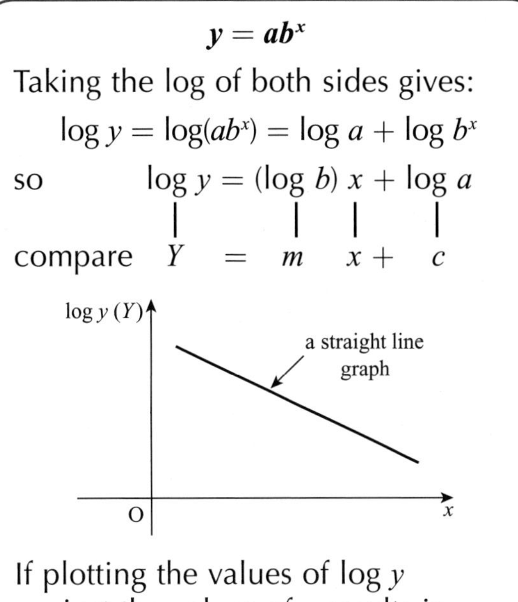


If plotting the values of log $y$ against the values of $x$ results in a straight line graph then this confirms a relationship of the form

$$y = ab^x$$

for suitable values of constants $a$ and $b$.

## Example

It is suspected that the relationship between $x$ and $y$ is of the form

$$y = ax^b$$

and the graph confirms this. Find the values of constants $a$ and $b$.

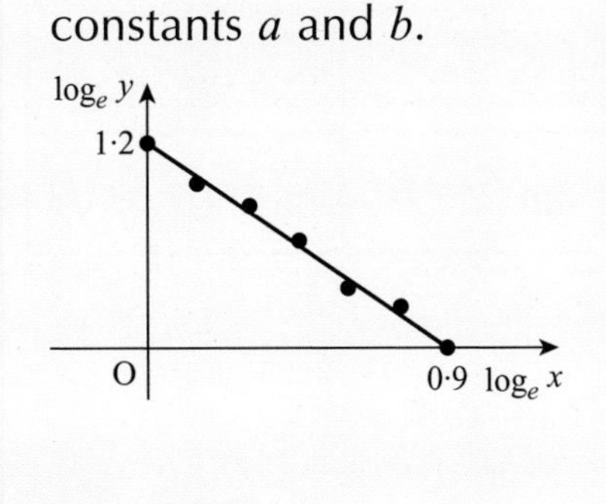


## Solution

$y = ax^b$ gives $\log_e y = \log_e(ax^b) = \log_e a + \log_e x^b$

so $\log_e y = b \log_e x + \log_e a$

compare $Y = mX + c$

For the given graph, the gradient ($m$) gives $b$.
Use $A(0, 1{\cdot}2)$ and $B(0{\cdot}9, 0)$

So $m_{AB} = \frac{1{\cdot}2 - 0}{0 - 0{\cdot}9} = \frac{1{\cdot}2}{-0{\cdot}9} = -\frac{12}{9} = -\frac{4}{3}$ So $b = -\frac{4}{3}$

The $y$-intercept $(0, c)$ gives $\log_e a$ ($c = \log_e a$).
The intercept is $(0, 1{\cdot}2)$
so $\log_e a = 1{\cdot}2 \Rightarrow a = e^{1{\cdot}2} \doteqdot 3{\cdot}32$ (using $e^x$ button)

This gives the relationship $\mathbf{y = 3{\cdot}32x^{-\frac{4}{3}}}$

# Using radians

## What is a radian?

Lay a radius-length along the circumference of any circle. Then the angle formed at the centre is 1 radian.

Since circumference $= 2\pi r$ then semicircle $= \pi r$ So $\pi$ radius-lengths make up a semicircle.

Looking at the angles formed at the centre

**$\pi$ radians = 180°**

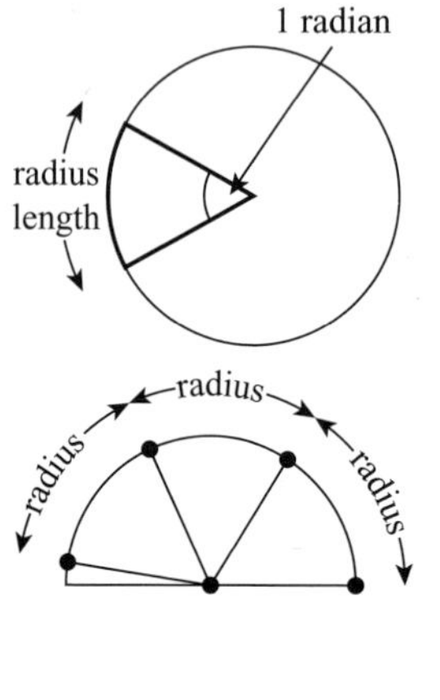


Some useful conversions:

| radians | | degrees |
|---|---|---|
| $\pi$ | ⟷ | 180° |
| $\frac{\pi}{6}$ | ⟷ | 30° |
| $\frac{\pi}{4}$ | ⟷ | 45° |
| $\frac{\pi}{3}$ | ⟷ | 60° |
| $\frac{\pi}{2}$ | ⟷ | 90° |
| $\frac{3\pi}{2}$ | ⟷ | 270° |
| $2\pi$ | ⟷ | 360° |

## Some exact values

For $\frac{\pi}{4}$ or 45° use half a square of side 1:

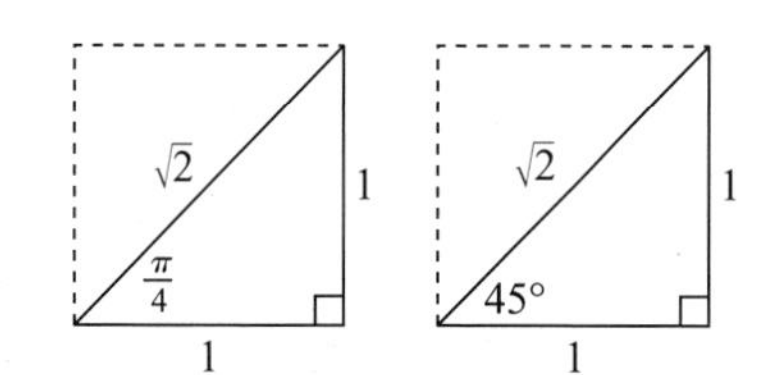


For $\frac{\pi}{6}$, $\frac{\pi}{3}$ or 30°, 60° use half an equilateral triangle of side 2:

Now use 'SOHCAHTOA' to find **exact** values:

| | | | | | |
|---|---|---|---|---|---|
| $\sin\frac{\pi}{4} = \frac{1}{\sqrt{2}}$ | $\sin 45° = \frac{1}{\sqrt{2}}$ | $\sin\frac{\pi}{6} = \frac{1}{2}$ | $\sin 30° = \frac{1}{2}$ | $\sin\frac{\pi}{3} = \frac{\sqrt{3}}{2}$ | $\sin 60° = \frac{\sqrt{3}}{2}$ |
| $\cos\frac{\pi}{4} = \frac{1}{\sqrt{2}}$ | $\cos 45° = \frac{1}{\sqrt{2}}$ | $\cos\frac{\pi}{6} = \frac{\sqrt{3}}{2}$ | $\cos 30° = \frac{\sqrt{3}}{2}$ | $\cos\frac{\pi}{3} = \frac{1}{2}$ | $\cos 60° = \frac{1}{2}$ |
| $\tan\frac{\pi}{4} = 1$ | $\tan 45° = 1$ | $\tan\frac{\pi}{6} = \frac{1}{\sqrt{3}}$ | $\tan 30° = \frac{1}{\sqrt{3}}$ | $\tan\frac{\pi}{3} = \sqrt{3}$ | $\tan 60° = \sqrt{3}$ |

**Example 1**

Find the exact value of $1 - \sin^2\frac{\pi}{4}$

**Solution**

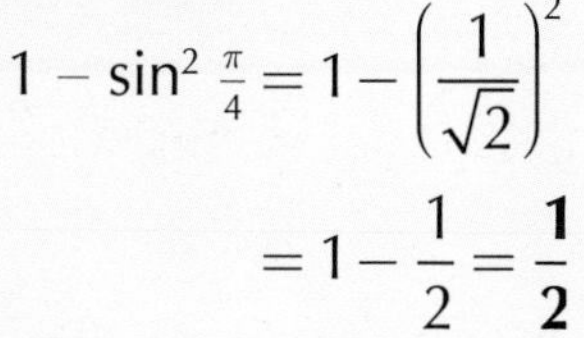

$$1 - \sin^2\frac{\pi}{4} = 1 - \left(\frac{1}{\sqrt{2}}\right)^2$$

$$= 1 - \frac{1}{2} = \mathbf{\frac{1}{2}}$$

**Example 2**

Solve $\sin x = 0{\cdot}3$ for $0 \le x \le \frac{\pi}{2}$

**Solution**

Put calculator into 'Radian Mode' and enter 0·3. Now use $\sin^{-1}$ giving $\boldsymbol{x = 0{\cdot}305}$ (to 3 sig. figs) radians.

Know your own calculator!! How do you change from Degree mode (D or DEG) to Radian mode (R or RAD)?

## The trig graphs using radian measure

Reading values from the three trig graphs with the angles measured in radians gives:

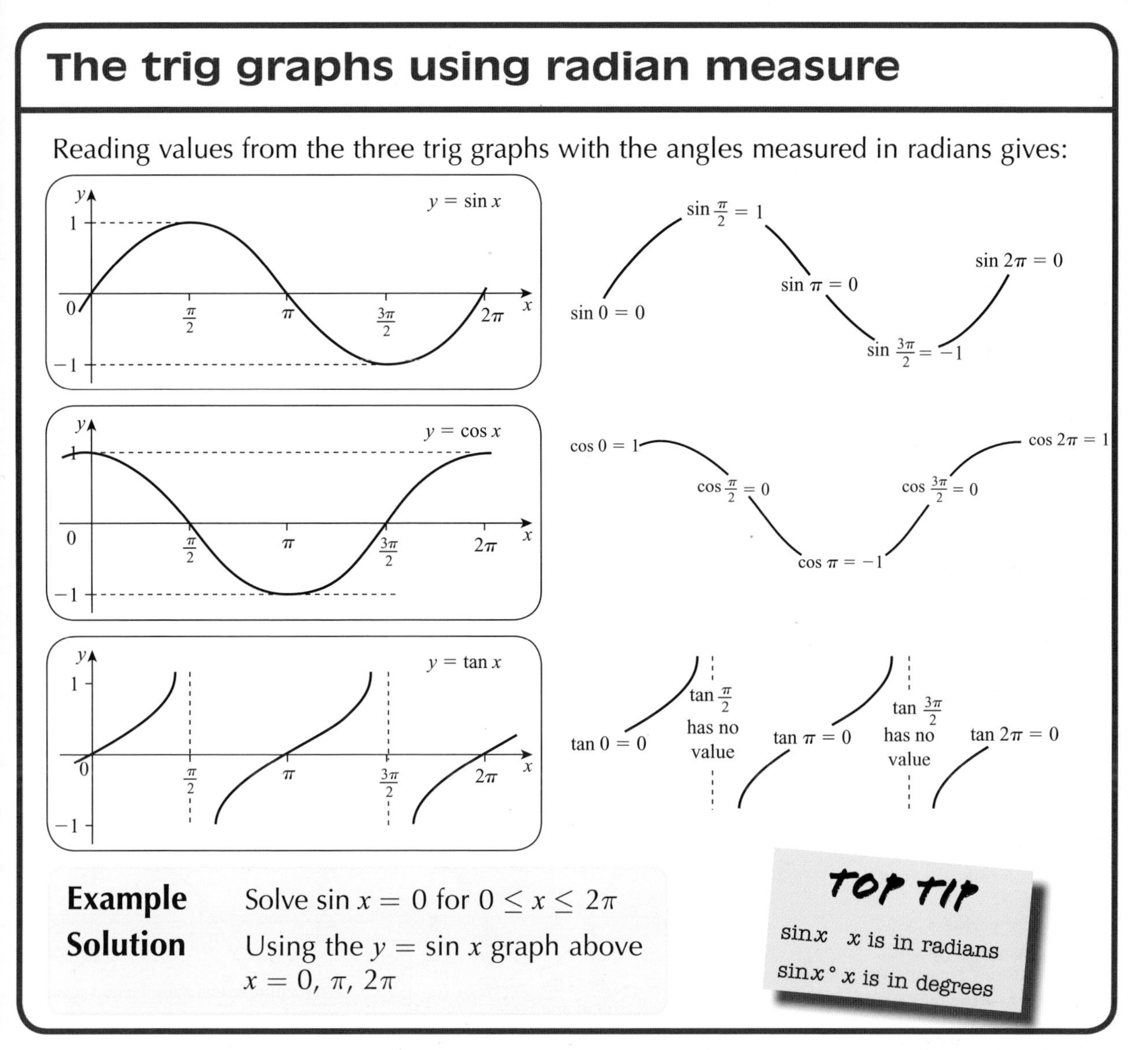


**Example** Solve $\sin x = 0$ for $0 \leq x \leq 2\pi$

**Solution** Using the $y = \sin x$ graph above
$x = 0, \pi, 2\pi$

**TOP TIP**

$\sin x$ $x$ is in radians

$\sin x°$ $x$ is in degrees

## Conversions

Other conversations can be done using "proportion":

| Radians | | Degrees | Degrees | | Radians |
|---|---|---|---|---|---|
| $\pi$ | ⟷ | 180 | 180 | ⟷ | $\pi$ |
| 1 | ⟷ | $\frac{180}{\pi}$ | 1 | ⟷ | $\frac{\pi}{180}$ |
| $x$ | ⟷ | $\frac{180}{\pi}x$ | $x$ | ⟷ | $\frac{\pi}{180}x$ |

# Quick Test 8

1. Find the exact value of $3 - 2\cos^2\left(\frac{\pi}{6}\right)$
2. Solve $\cos x = 0{\cdot}8$ for $0 \leq x \leq \frac{\pi}{2}$ giving your answer to 2 decimal places.
3. Solve $\cos x = 1$ for $0 \leq x \leq 2\pi$
4. Change
   a) $\frac{7\pi}{4}$ into degrees
   b) 40° into radians.

# Related trig graphs

## Graphs related to $y = \sin x$

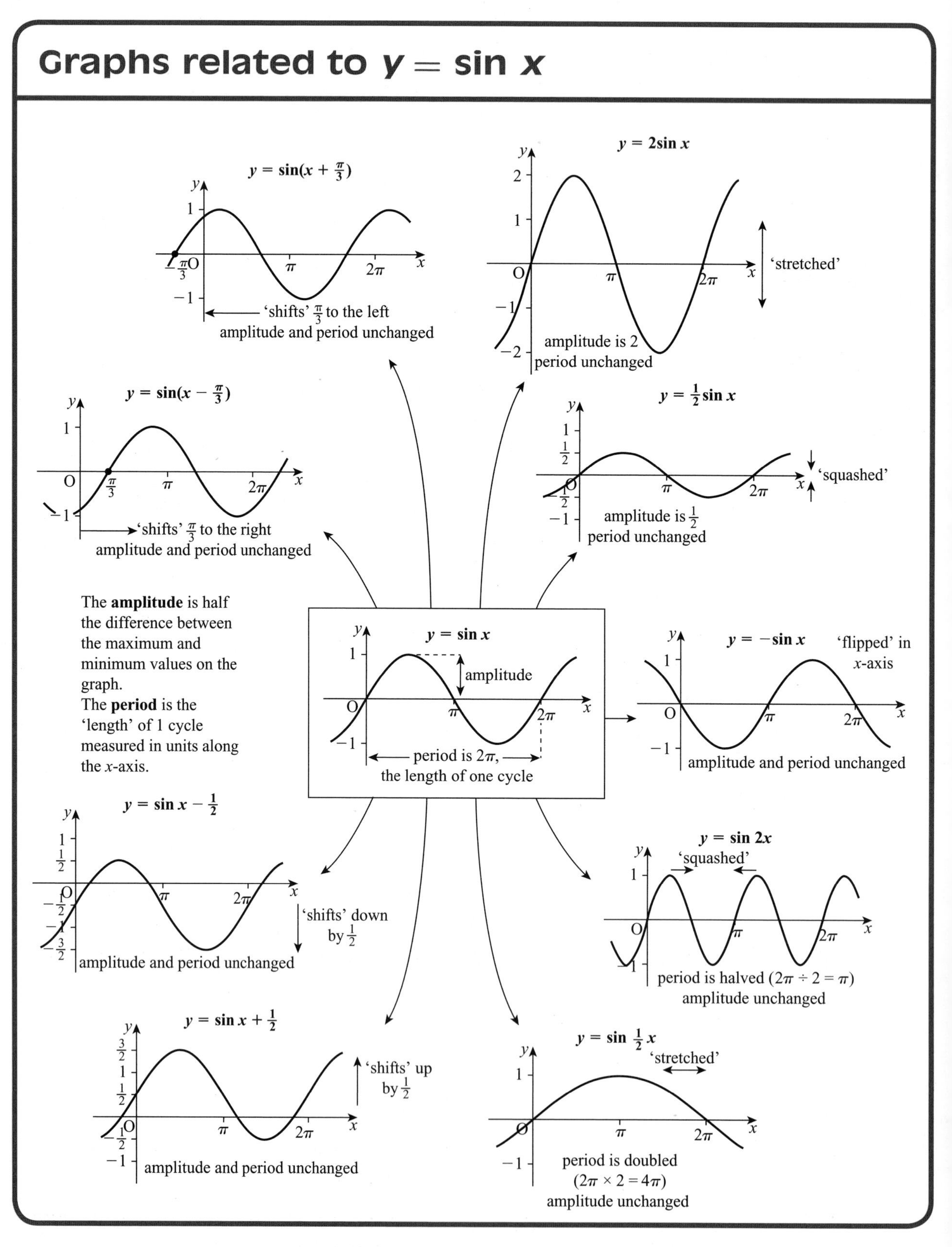

## Summary of effects

Here is a description of the effect to the graphs $y = \cos x$ or $y = \sin x$

$$y = a\cos(bx \pm c) \pm d$$
$$y = a\sin(bx \pm c) \pm d$$

- $a$: If $a > 0$ the amplitude is $a$. If $a < 0$ the graph 'flips' in the $x$-axis. The amplitude is the magnitude of $a$ (i.e. ignore the negative sign).
- $b$: Alters the period to $\frac{2\pi}{b}$ $(b > 0)$
- $c$: 'Shifts' the graph by $c$ units left for $+c$ or right for $-c$
- $d$: 'Shifts' the graph by $d$ units up for $+d$ or down for $-d$

**TOP TIP**

For the left/right up/down 'shifts' $c$ and $d$ are positive.

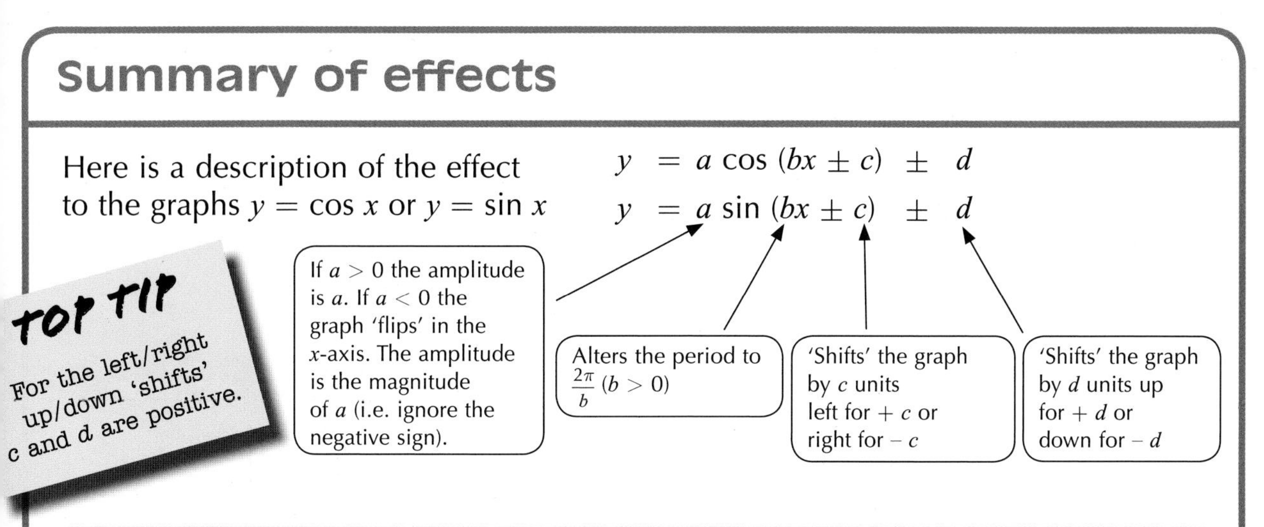

### Example

The diagram shows part of the graph of $y = a\sin(x - b) + c$. Find the values of $a$, $b$ and $c$.

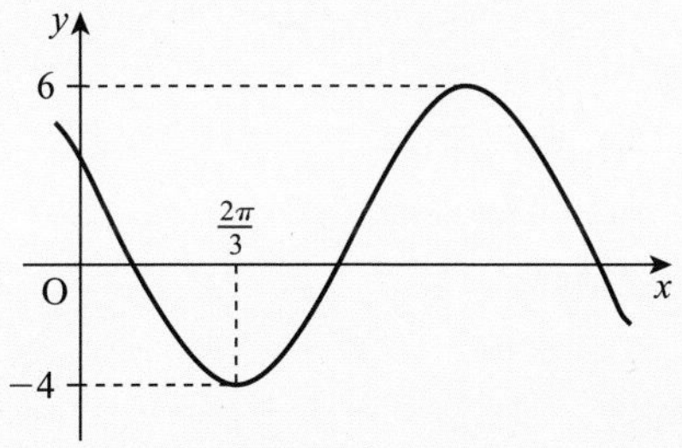


### Solution

Compared with $y = \sin x$... The amplitude is 5 (half the difference between –4 and 6). However the graph is 'flipped' in the $x$-axis so $\boldsymbol{a = -5}$. Also the graph has moved up 1 (max/min 6 and –4 instead of 5 and –5) so $\boldsymbol{c = 1}$.

It has also moved to the right: $\frac{2\pi}{3}$ instead of $\frac{\pi}{2}$

...now $\frac{2\pi}{3} - \frac{\pi}{2} = \frac{4\pi}{6} - \frac{3\pi}{6} = \frac{\pi}{6}$

so $\boldsymbol{b = \frac{\pi}{6}}$. The graph is $y = -5\sin(x - \frac{\pi}{6}) + 1$

## Sketching trig graphs

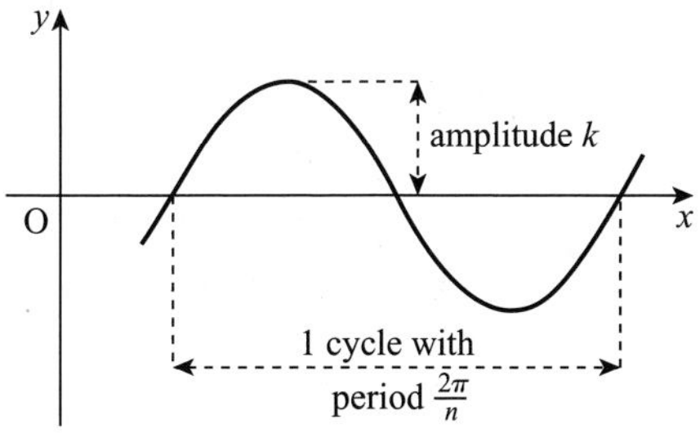


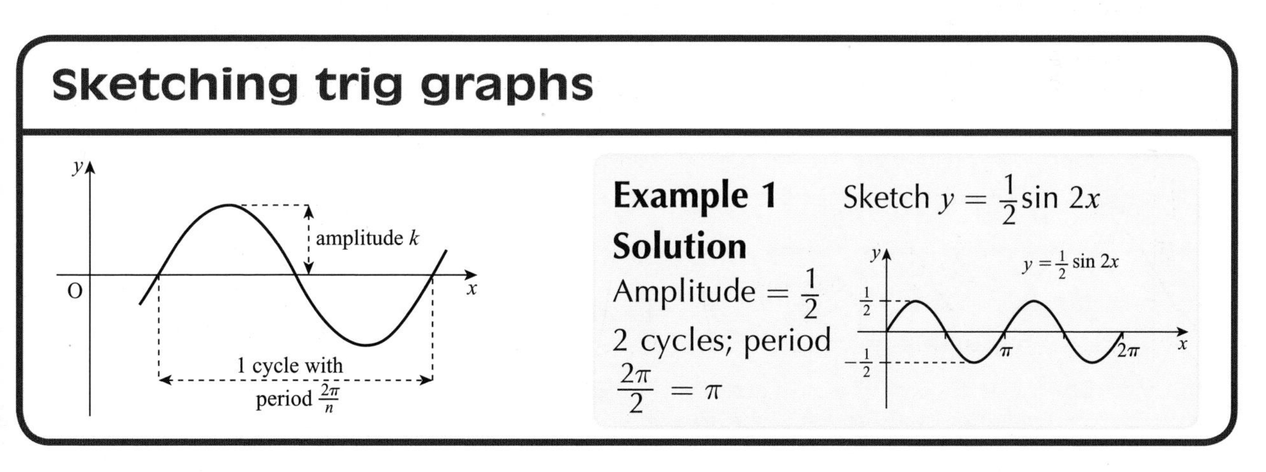

**Example 1** Sketch $y = \frac{1}{2}\sin 2x$

**Solution**

Amplitude $= \frac{1}{2}$

2 cycles; period $\frac{2\pi}{2} = \pi$

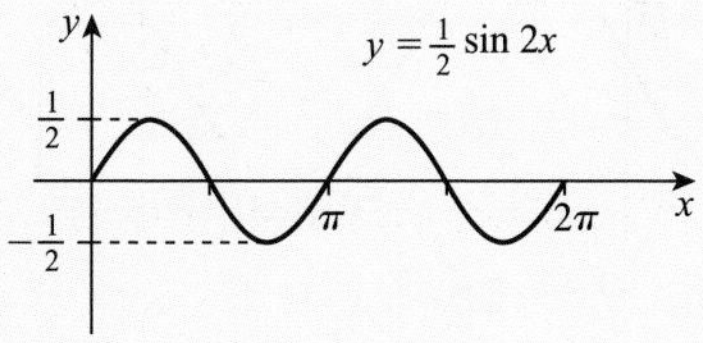


# Quick Test 9

1. Find the values of $a$ and $b$ for this graph:

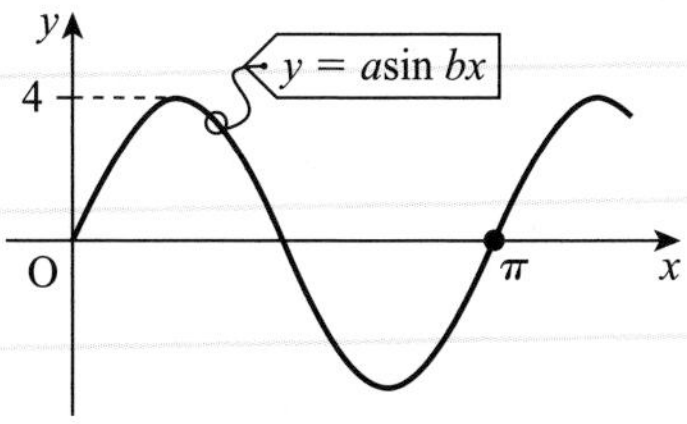


2. a) Sketch the graph $y = 5\cos\left(x - \frac{\pi}{4}\right)$ for $0 \le x \le 2\pi$

   b) Find the coordinates of the maximum and minimum points on your sketched graph.

# Problem solving using trig

## Working in 3D

The angle between a line and a plane:

If $QR$ is the **projection** or 'vertical shadow' of the line $PQ$ on the plane then

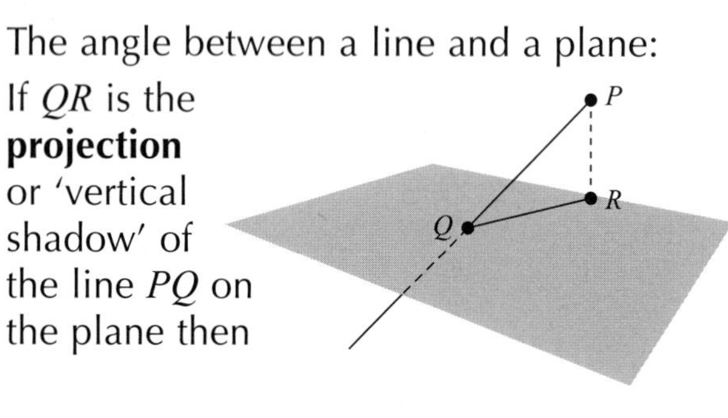


$\angle PQR$ is the angle between the line and plane.

The angle between two planes:

Choose a point $B$ on the line $l$ where the two planes meet. Draw $AB$ on one plane and $BC$ on the other, both perpendicular to line $l$.

$\angle ABC$ is the angle between the two planes.

### Example

A square-based pyramid has vertex $P$, 3 meters above $Q$, the centre of the base. The square base has side length 6 meters. Find the angle between the base and a sloping face in radians.

### Solution

Choose $M$, the midpoint of one of the sides of the square base. Required angle is $\angle QMP = \frac{\pi}{4}$ (isosceles right-angled triangle).

## Basic trig identities

$\frac{\sin\theta}{\tan\theta} = \tan\theta$ $\quad$ $\sin^2\theta + \cos^2\theta = 1$

rearranging gives

$\sin\theta = \tan\theta\cos\theta$ $\quad$ $\sin^2\theta = 1 - \cos^2\theta$

$\cos^2\theta = 1 - \sin^2\theta$

$\cos\theta = \frac{\sin\theta}{\tan\theta}$

**Example** Prove $\tan^2 A = \frac{1-\cos^2 A}{1-\sin^2 A}$

**Solution** $\tan^2 A = \tan A \times \tan A = \frac{\sin A}{\cos A} \times \frac{\sin A}{\cos A}$

$= \frac{\sin^2 A}{\cos^2 A} = \frac{1-\cos^2 A}{1-\sin^2 A}$

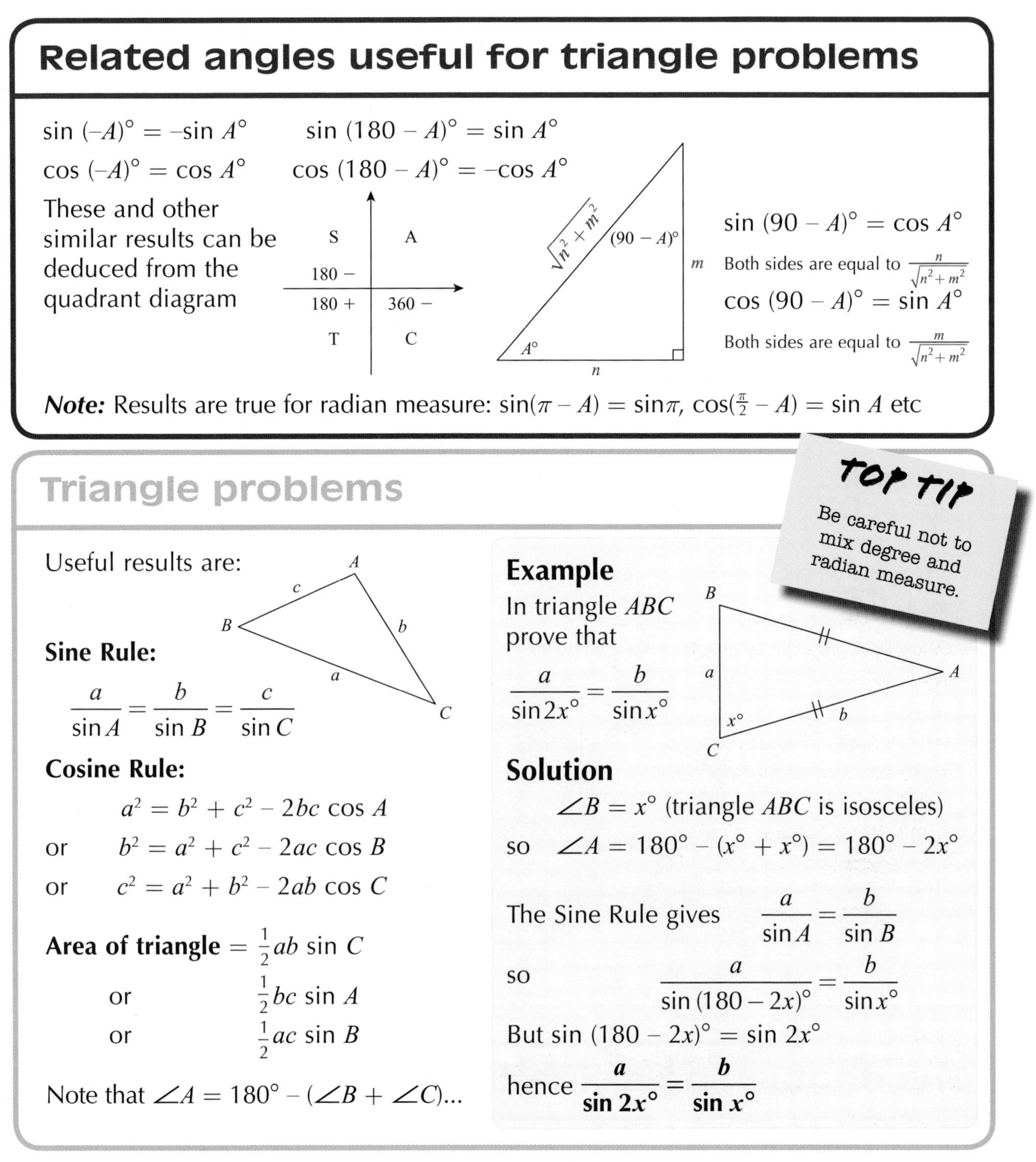

## Related angles useful for triangle problems

$\sin(-A)^\circ = -\sin A^\circ$ $\quad\quad \sin(180 - A)^\circ = \sin A^\circ$

$\cos(-A)^\circ = \cos A^\circ$ $\quad\quad \cos(180 - A)^\circ = -\cos A^\circ$

These and other similar results can be deduced from the quadrant diagram



$\sin(90 - A)^\circ = \cos A^\circ$

Both sides are equal to $\frac{n}{\sqrt{n^2+m^2}}$

$\cos(90 - A)^\circ = \sin A^\circ$

Both sides are equal to $\frac{m}{\sqrt{n^2+m^2}}$

***Note:*** Results are true for radian measure: $\sin(\pi - A) = \sin\pi$, $\cos(\frac{\pi}{2} - A) = \sin A$ etc

## Triangle problems

**TOP TIP**

Be careful not to mix degree and radian measure.

Useful results are:



**Sine Rule:**

$$\frac{a}{\sin A} = \frac{b}{\sin B} = \frac{c}{\sin C}$$

**Cosine Rule:**

$$a^2 = b^2 + c^2 - 2bc\cos A$$

or $\quad b^2 = a^2 + c^2 - 2ac\cos B$

or $\quad c^2 = a^2 + b^2 - 2ab\cos C$

**Area of triangle** $= \frac{1}{2}ab\sin C$

or $\quad \frac{1}{2}bc\sin A$

or $\quad \frac{1}{2}ac\sin B$

Note that $\angle A = 180^\circ - (\angle B + \angle C)$...

**Example**

In triangle $ABC$ prove that

$$\frac{a}{\sin 2x^\circ} = \frac{b}{\sin x^\circ}$$



**Solution**

$\angle B = x^\circ$ (triangle $ABC$ is isosceles)

so $\quad \angle A = 180^\circ - (x^\circ + x^\circ) = 180^\circ - 2x^\circ$

The Sine Rule gives $\quad \frac{a}{\sin A} = \frac{b}{\sin B}$

so $\quad \frac{a}{\sin(180 - 2x)^\circ} = \frac{b}{\sin x^\circ}$

But $\sin(180 - 2x)^\circ = \sin 2x^\circ$

hence $\quad \boldsymbol{\frac{a}{\sin 2x^\circ} = \frac{b}{\sin x^\circ}}$

# Quick Test 10

**1.** A billboard is supported by a wooden support wedge as shown in the diagram. The wedge has dimensions as shown in the diagram on the right.

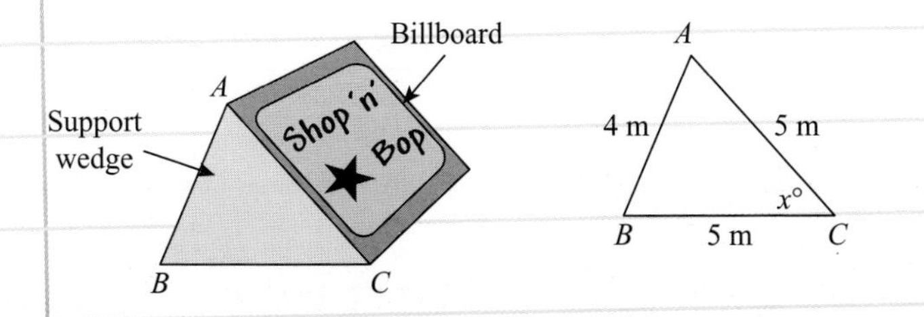


Show that $\cos\frac{1}{2}x = \frac{5}{4}\sin x$ where $x^\circ$ is the angle at which the billboard is inclined to the horizontal.

Hint: use the sine rule.

# Trig formulae

## The addition formulae

The following are the 'addition formulae' and are true for all values of $A$ and $B$:

$$\sin(A + B) = \sin A \cos B + \cos A \sin B$$
$$\sin(A - B) = \sin A \cos B - \cos A \sin B$$
$$\cos(A + B) = \cos A \cos B - \sin A \sin B$$
$$\cos(A - B) = \cos A \cos B + \sin A \sin B$$

### Example 1

Show that $\cos(\frac{\pi}{2} + x) = -\sin x$

### Solution

$$\cos(\tfrac{\pi}{2} + x) = \cos\tfrac{\pi}{2}\cos x - \sin\tfrac{\pi}{2}\sin x$$
$$= 0 \times \cos x - 1 \times \sin x = -\sin x$$

### Example 2

If $A$ and $B$ are acute angles with $\sin A = \frac{3}{5}$ and $\cos B = \frac{12}{13}$ find the **exact** value of $\cos(A - B)$.

### Solution

Draw right-angled triangles showing $\angle A$ and $\angle B$ (use Pythagoras' Theorem to find third side).

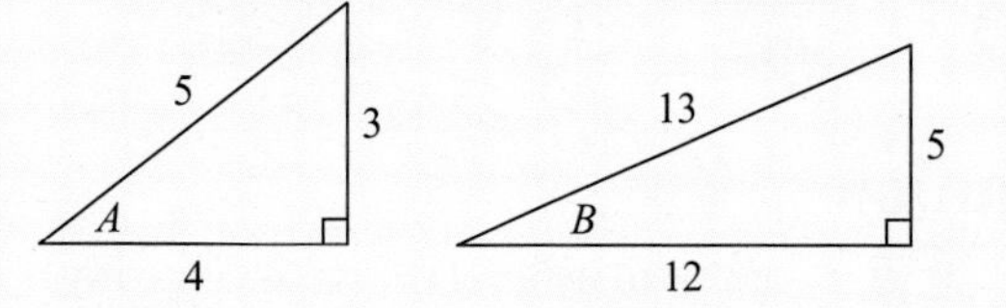


Then $\cos(A - B) = \cos A \cos B + \sin A \sin B$

$$= \frac{4}{5} \times \frac{12}{13} + \frac{3}{5} \times \frac{5}{13}$$
$$= \frac{48}{65} + \frac{15}{65} = \mathbf{\frac{63}{65}}$$

### Example 3

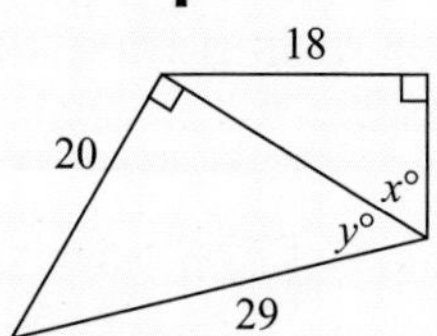


Show that the exact value of $\sin(x + y)°$ is $\frac{378 + 60\sqrt{13}}{609}$

### Solution

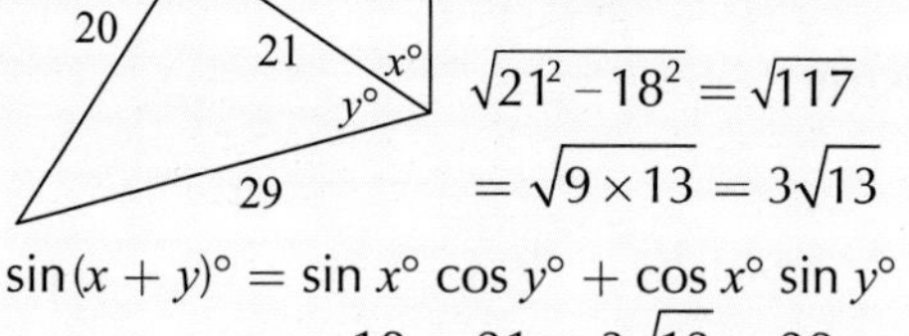


Pythagoras' Theorem is used to find the two missing lengths

$$\sqrt{21^2 - 18^2} = \sqrt{117}$$
$$= \sqrt{9 \times 13} = 3\sqrt{13}$$

$$\sin(x + y)° = \sin x° \cos y° + \cos x° \sin y°$$
$$= \frac{18}{21} \times \frac{21}{29} + \frac{3\sqrt{13}}{21} \times \frac{20}{29}$$
$$= \frac{18 \times 21}{21 \times 29} + \frac{3\sqrt{13} \times 20}{21 \times 29}$$
$$= \frac{378 + 60\sqrt{13}}{609}$$

## The double angle sine formula

The double angle formula is:

$$\sin 2A = 2\sin A \cos A$$

Examples using this as a template are:

$$\sin 4\theta = 2\sin 2\theta \cos 2\theta$$
$$\sin P = 2\sin\frac{P}{2}\cos\frac{P}{2}$$

**Example** Show that $(\cos x + \sin x)^2 = 1 + \sin 2x$

### Solution

$$(\cos x + \sin x)^2 = (\cos x + \sin x)(\cos x + \sin x)$$
$$= \cos^2 x + \cos x \sin x + \sin x \cos x + \sin^2 x$$
$$= \underbrace{\cos^2 x + \sin^2 x}_{1} + \underbrace{2\sin x \cos x}_{\sin 2x}$$
$$= 1 + \sin 2x$$

# The double angle cosine formulae

There are three versions of this double angle formula:

$\cos 2A = 2\cos^2 A - 1$
or $\cos^2 A - \sin^2 A$
or $1 - 2\sin^2 A$

Examples using these as a template are:

$$\cos\theta = 2\cos^2\frac{\theta}{2} - 1 = 1 - 2\sin^2\frac{\theta}{2}$$

Here are two useful rearrangements:

$$\cos^2 A = \frac{1}{2}(1 + \cos 2A)$$

$$\sin^2 A = \frac{1}{2}(1 - \cos 2A)$$

$$\cos^2\frac{P}{2} = \frac{1}{2}(1 + \cos P)$$

**TOP TIP**

The addition formulae and the double angle formulae will be given to you during your exam. But you should memorise them.

### Example 1

Show that $\sin\frac{\pi}{12} = \sqrt{\frac{2-\sqrt{3}}{4}}$ Hint: use $\sin^2 A = \frac{1}{2}(1 - \cos 2A)$ with $A = \frac{\pi}{12}$

### Solution

$$\sin^2\frac{\pi}{12} = \frac{1}{2}\left(1 - \cos\left(2 \times \frac{\pi}{12}\right)\right) = \frac{1}{2}\left(1 - \cos\frac{\pi}{6}\right) = \frac{1}{2}\left(1 - \frac{\sqrt{3}}{2}\right)$$

$$= \frac{1}{2}\left(\frac{2}{2} - \frac{\sqrt{3}}{2}\right) = \frac{1}{2}\left(\frac{2-\sqrt{3}}{2}\right) = \frac{2-\sqrt{3}}{4}$$

So $\sin\frac{\pi}{12} = \sqrt{\frac{2-\sqrt{3}}{4}}$

### Example 2

If $\tan\alpha = \frac{\sqrt{7}}{2}$ find the **exact** value of $\cos 2\alpha$ where $0 \leq \alpha \leq \frac{\pi}{2}$

### Solution

Here is a right-angled triangle for which $\tan\alpha = \frac{\sqrt{7}}{2}$:

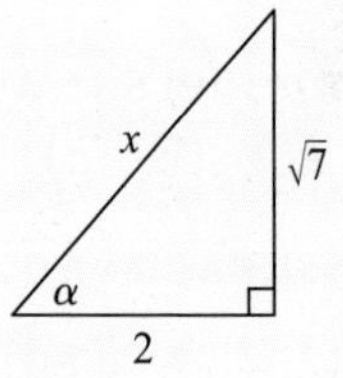


$$x^2 = 2^2 + (\sqrt{7})^2 = 4 + 7 = 11$$

So $x = \sqrt{11}$

$$\cos 2\alpha = 2\cos^2\alpha - 1 = 2 \times \frac{2}{\sqrt{11}} \times \frac{2}{\sqrt{11}} - 1 = \frac{8}{11} - 1 = \frac{8}{11} - \frac{11}{11} = -\frac{3}{11}$$

**Note:** Other versions of $\cos 2\alpha$ e.g. $1 - 2\sin^2\alpha$ could have been used.

# Quick Test 11

1. Show that $(\cos\theta + \sin\theta)(\cos\theta - \sin\theta) = \cos 2\theta$
2. If $0 < \alpha < \frac{\pi}{2}$ find the exact value of $\sin 2\alpha$ given that $\tan\alpha = \frac{1}{3}$
3. 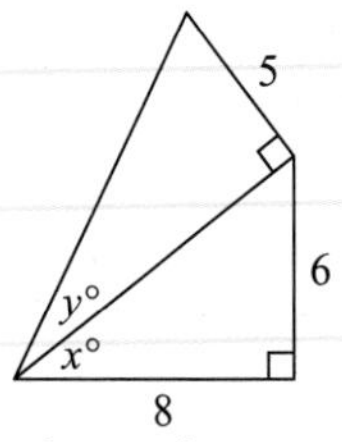


   Show that $\cos(x + y)^\circ = \frac{1}{\sqrt{5}}$

# The wave function

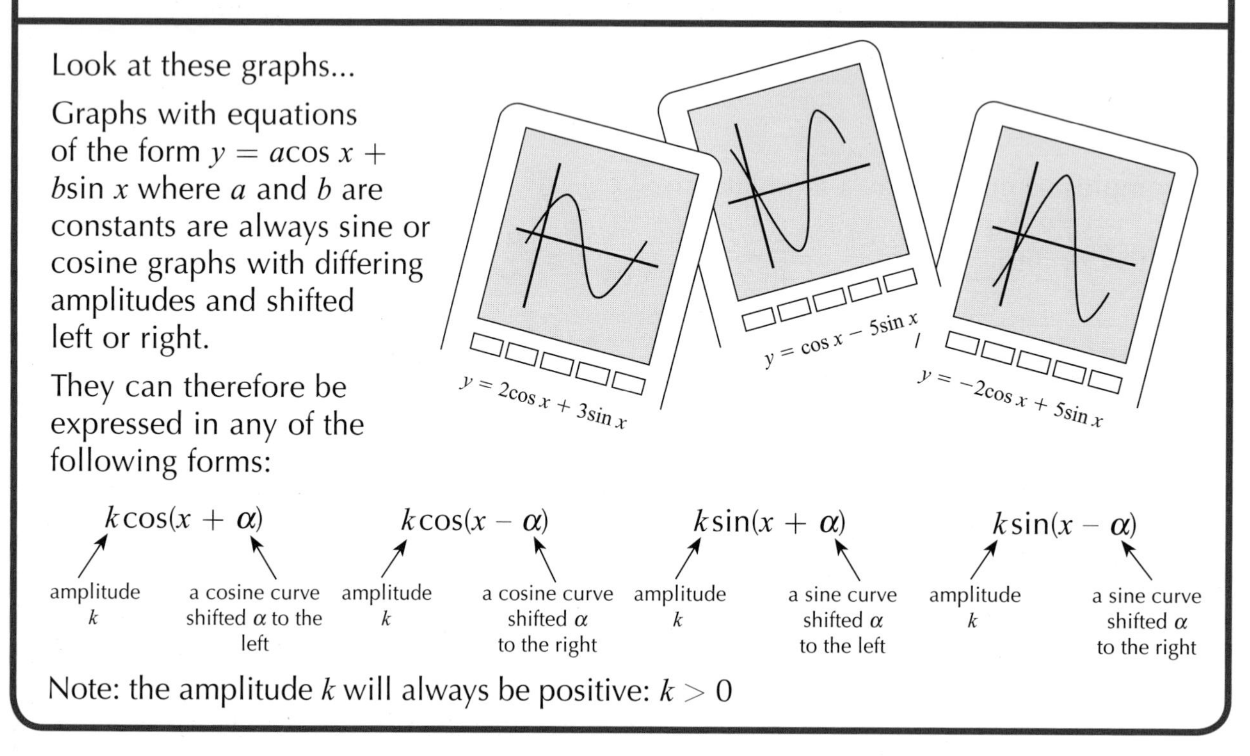


## Linear combinations of sin x and cos x

Look at these graphs...

Graphs with equations of the form $y = a\cos x + b\sin x$ where $a$ and $b$ are constants are always sine or cosine graphs with differing amplitudes and shifted left or right.

They can therefore be expressed in any of the following forms:

- $k\cos(x + \alpha)$ — amplitude $k$; a cosine curve shifted $\alpha$ to the left
- $k\cos(x - \alpha)$ — amplitude $k$; a cosine curve shifted $\alpha$ to the right
- $k\sin(x + \alpha)$ — amplitude $k$; a sine curve shifted $\alpha$ to the left
- $k\sin(x - \alpha)$ — amplitude $k$; a sine curve shifted $\alpha$ to the right

Note: the amplitude $k$ will always be positive: $k > 0$

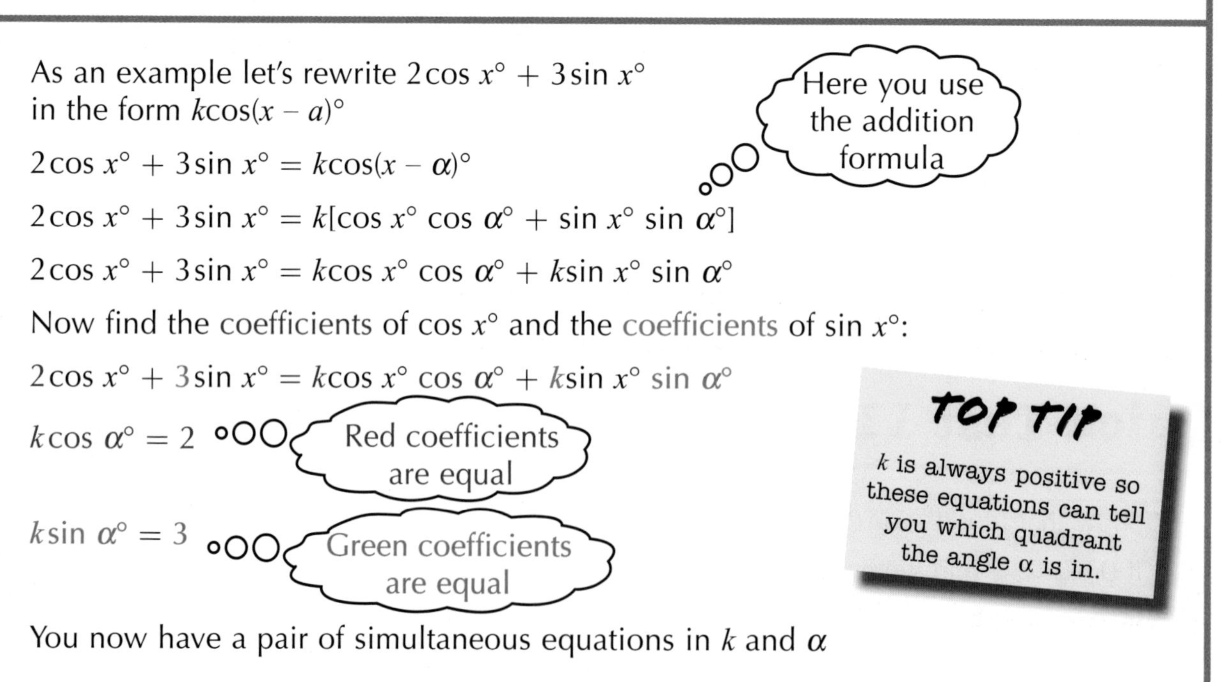

## Comparing coefficients

As an example let's rewrite $2\cos x^\circ + 3\sin x^\circ$ in the form $k\cos(x - a)^\circ$

*Here you use the addition formula*

$2\cos x^\circ + 3\sin x^\circ = k\cos(x - \alpha)^\circ$

$2\cos x^\circ + 3\sin x^\circ = k[\cos x^\circ \cos \alpha^\circ + \sin x^\circ \sin \alpha^\circ]$

$2\cos x^\circ + 3\sin x^\circ = k\cos x^\circ \cos \alpha^\circ + k\sin x^\circ \sin \alpha^\circ$

Now find the coefficients of cos $x^\circ$ and the coefficients of sin $x^\circ$:

$2\cos x^\circ + 3\sin x^\circ = k\cos x^\circ \cos \alpha^\circ + k\sin x^\circ \sin \alpha^\circ$

$k\cos \alpha^\circ = 2$ — *Red coefficients are equal*

$k\sin \alpha^\circ = 3$ — *Green coefficients are equal*

You now have a pair of simultaneous equations in $k$ and $\alpha$

**TOP TIP**

$k$ is always positive so these equations can tell you which quadrant the angle $\alpha$ is in.

## Determining $k$ and $\alpha$

$\left.\begin{array}{l} k\cos\alpha° = 2 \\ k\sin\alpha° = 3 \end{array}\right\}$ $k$ is always positive so these equations tell you that $\sin\alpha°$ and $\cos\alpha°$ are both positive. This means $\alpha°$ is an angle in the 1st quadrant so $0° < \alpha° < 90°$

Divide both sides and use: $\frac{\sin\alpha}{\cos\alpha} = \tan\alpha$

$$\frac{\cancel{k}\sin\alpha°}{\cancel{k}\cos\alpha°} = \frac{3}{2} \Rightarrow \tan\alpha° = 1{\cdot}5 \Rightarrow \alpha° \doteqdot 56{\cdot}3° \text{ (1st quadrant only)}$$

Square both sides and add then use: $\sin^2\alpha + \cos^2\alpha = 1$

$$(k\sin\alpha°)^2 + (k\cos\alpha°)^2 = 3^2 + 2^2$$
$$\Rightarrow k^2\sin^2\alpha° + k^2\cos^2\alpha° = 9 + 4$$
$$\Rightarrow k^2(\sin^2\alpha° + \cos^2\alpha°) = 13$$
$$\Rightarrow k^2 \times 1 = 13 \Rightarrow k^2 = 13 \Rightarrow k = \sqrt{13} \text{ (positive)}$$

Here is a radian example:

$\left.\begin{array}{l} k\sin\alpha = 2 \\ k\cos\alpha = -5 \end{array}\right\}$ $k$ is always positive so these equations tell you that $\sin\alpha$ is positive and $\cos\alpha$ is negative. This means $\alpha$ is an angle in the 2nd quadrant so $\frac{\pi}{2} < \alpha < \pi$

Divide both sides and use: $\frac{\sin\alpha}{\cos\alpha} = \tan\alpha$

$$\frac{\cancel{k}\sin\alpha}{\cancel{k}\cos\alpha} = \frac{2}{-5} \Rightarrow \tan\alpha = -\frac{2}{5} \Rightarrow \tan\alpha = -0{\cdot}4$$
$$\alpha = \pi - 0{\cdot}380...$$
$$= 2{\cdot}761... \doteqdot 2{\cdot}76 \text{ (2nd quadrant)}$$

Using $\tan\alpha = 0{\cdot}4$ the 1st quadrant angle is $0{\cdot}380...$ (radians)

Square both sides and add then use: $\sin^2\alpha + \cos^2\alpha = 1$

$$(k\sin\alpha)^2 + (k\cos\alpha)^2 = 2^2 + (-5)^2$$
$$\Rightarrow k^2\sin^2\alpha + k^2\cos^2\alpha = 4 + 25$$
$$\Rightarrow k^2(\sin^2\alpha + \cos^2\alpha) = 29$$
$$\Rightarrow k^2 \times 1 = 29 \Rightarrow k^2 = 29 \Rightarrow k = \sqrt{29} \text{ (positive)}$$

## The final form

**Example** Express $2\cos x° + 3\sin x°$ in the form $k\cos(x - \alpha)°$ where $k > 0$

**Solution** From above $k = \sqrt{13}$ and $\alpha \doteqdot 56{\cdot}3$

so $2\cos x° + 3\sin x° \doteqdot \sqrt{13}\cos(x - 56{\cdot}3)°$

## Quick Test 12

1. Express $\cos x° - 3\sin x°$ in the form $k\cos(x + \alpha)°$ where $k > 0$ and $0 \leq \alpha < 360$
2. Express $-\cos x - \sqrt{3}\sin x$ in the form $k\sin(x + \alpha)$ where $k > 0$ and $0 \leq \alpha < 2\pi$

# Vectors: basic properties

## What is a vector?

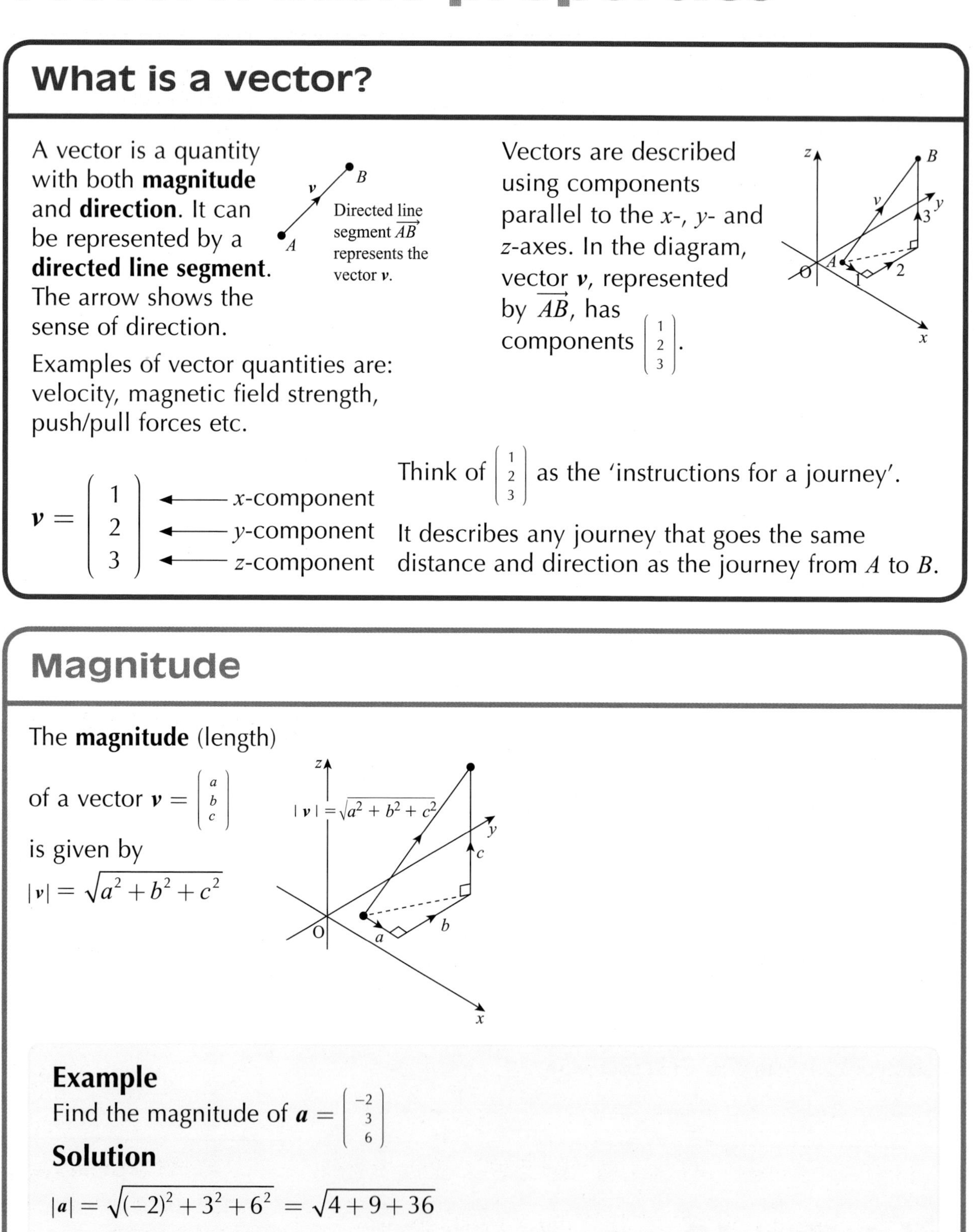

A vector is a quantity with both **magnitude** and **direction**. It can be represented by a **directed line segment**. The arrow shows the sense of direction.

Directed line segment $\overrightarrow{AB}$ represents the vector $\boldsymbol{v}$.

Examples of vector quantities are: velocity, magnetic field strength, push/pull forces etc.

Vectors are described using components parallel to the $x$-, $y$- and $z$-axes. In the diagram, vector $\boldsymbol{v}$, represented by $\overrightarrow{AB}$, has components $\begin{pmatrix} 1 \\ 2 \\ 3 \end{pmatrix}$.

$$\boldsymbol{v} = \begin{pmatrix} 1 \\ 2 \\ 3 \end{pmatrix} \begin{matrix} \leftarrow x\text{-component} \\ \leftarrow y\text{-component} \\ \leftarrow z\text{-component} \end{matrix}$$

Think of $\begin{pmatrix} 1 \\ 2 \\ 3 \end{pmatrix}$ as the 'instructions for a journey'.

It describes any journey that goes the same distance and direction as the journey from $A$ to $B$.

## Magnitude

The **magnitude** (length) of a vector $\boldsymbol{v} = \begin{pmatrix} a \\ b \\ c \end{pmatrix}$ is given by

$$|\boldsymbol{v}| = \sqrt{a^2 + b^2 + c^2}$$

### Example

Find the magnitude of $\boldsymbol{a} = \begin{pmatrix} -2 \\ 3 \\ 6 \end{pmatrix}$

### Solution

$$|\boldsymbol{a}| = \sqrt{(-2)^2 + 3^2 + 6^2} = \sqrt{4 + 9 + 36}$$

$$= \sqrt{49}$$

$$= 7$$

## Equal vectors

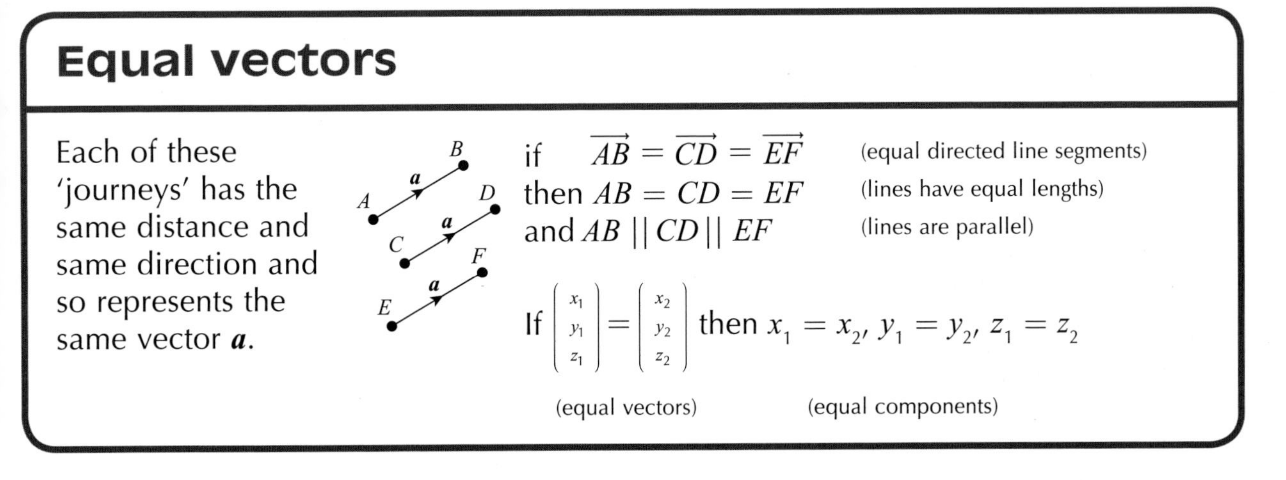


Each of these 'journeys' has the same distance and same direction and so represents the same vector ***a***.

if $\overrightarrow{AB} = \overrightarrow{CD} = \overrightarrow{EF}$ (equal directed line segments)
then $AB = CD = EF$ (lines have equal lengths)
and $AB \,||\, CD \,||\, EF$ (lines are parallel)

If $\begin{pmatrix} x_1 \\ y_1 \\ z_1 \end{pmatrix} = \begin{pmatrix} x_2 \\ y_2 \\ z_2 \end{pmatrix}$ then $x_1 = x_2$, $y_1 = y_2$, $z_1 = z_2$

(equal vectors) (equal components)

## The zero vector

$\mathbf{0} = \begin{pmatrix} 0 \\ 0 \\ 0 \end{pmatrix}$ The zero vector has magnitude $|\mathbf{0}| = 0$ but has **no** direction defined.

## Addition

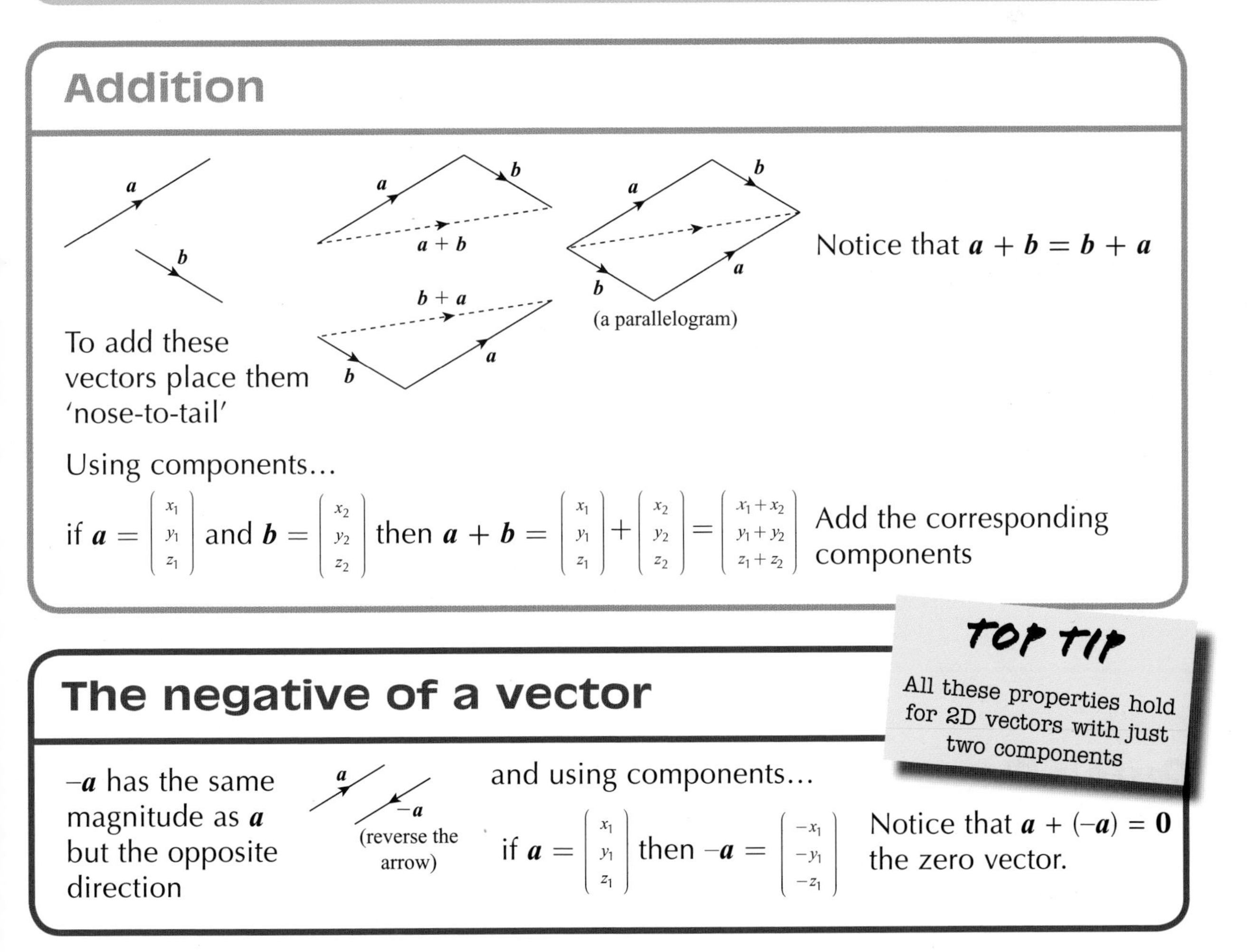


(a parallelogram)

Notice that $\boldsymbol{a} + \boldsymbol{b} = \boldsymbol{b} + \boldsymbol{a}$

To add these vectors place them 'nose-to-tail'

Using components…

if $\boldsymbol{a} = \begin{pmatrix} x_1 \\ y_1 \\ z_1 \end{pmatrix}$ and $\boldsymbol{b} = \begin{pmatrix} x_2 \\ y_2 \\ z_2 \end{pmatrix}$ then $\boldsymbol{a} + \boldsymbol{b} = \begin{pmatrix} x_1 \\ y_1 \\ z_1 \end{pmatrix} + \begin{pmatrix} x_2 \\ y_2 \\ z_2 \end{pmatrix} = \begin{pmatrix} x_1 + x_2 \\ y_1 + y_2 \\ z_1 + z_2 \end{pmatrix}$ Add the corresponding components

**TOP TIP**

All these properties hold for 2D vectors with just two components

## The negative of a vector

$-\boldsymbol{a}$ has the same magnitude as ***a*** but the opposite direction



(reverse the arrow)

and using components…

if $\boldsymbol{a} = \begin{pmatrix} x_1 \\ y_1 \\ z_1 \end{pmatrix}$ then $-\boldsymbol{a} = \begin{pmatrix} -x_1 \\ -y_1 \\ -z_1 \end{pmatrix}$

Notice that $\boldsymbol{a} + (-\boldsymbol{a}) = \mathbf{0}$ the zero vector.

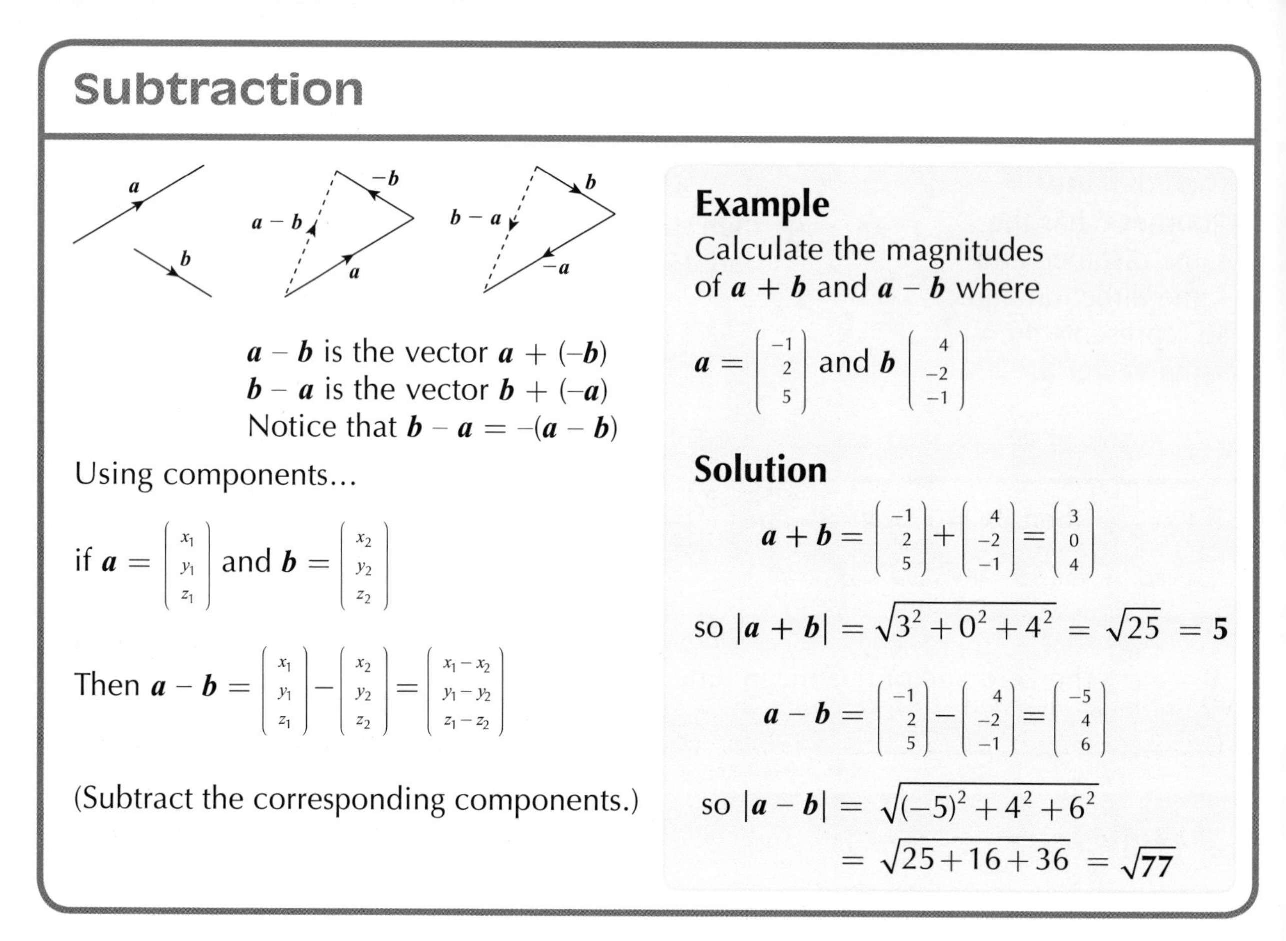

## Subtraction



$\boldsymbol{a} - \boldsymbol{b}$ is the vector $\boldsymbol{a} + (-\boldsymbol{b})$

$\boldsymbol{b} - \boldsymbol{a}$ is the vector $\boldsymbol{b} + (-\boldsymbol{a})$

Notice that $\boldsymbol{b} - \boldsymbol{a} = -(\boldsymbol{a} - \boldsymbol{b})$

Using components...

if $\boldsymbol{a} = \begin{pmatrix} x_1 \\ y_1 \\ z_1 \end{pmatrix}$ and $\boldsymbol{b} = \begin{pmatrix} x_2 \\ y_2 \\ z_2 \end{pmatrix}$

Then $\boldsymbol{a} - \boldsymbol{b} = \begin{pmatrix} x_1 \\ y_1 \\ z_1 \end{pmatrix} - \begin{pmatrix} x_2 \\ y_2 \\ z_2 \end{pmatrix} = \begin{pmatrix} x_1 - x_2 \\ y_1 - y_2 \\ z_1 - z_2 \end{pmatrix}$

(Subtract the corresponding components.)

### Example

Calculate the magnitudes of $\boldsymbol{a} + \boldsymbol{b}$ and $\boldsymbol{a} - \boldsymbol{b}$ where

$\boldsymbol{a} = \begin{pmatrix} -1 \\ 2 \\ 5 \end{pmatrix}$ and $\boldsymbol{b} \begin{pmatrix} 4 \\ -2 \\ -1 \end{pmatrix}$

### Solution

$$\boldsymbol{a} + \boldsymbol{b} = \begin{pmatrix} -1 \\ 2 \\ 5 \end{pmatrix} + \begin{pmatrix} 4 \\ -2 \\ -1 \end{pmatrix} = \begin{pmatrix} 3 \\ 0 \\ 4 \end{pmatrix}$$

so $|\boldsymbol{a} + \boldsymbol{b}| = \sqrt{3^2 + 0^2 + 4^2} = \sqrt{25} = \mathbf{5}$

$$\boldsymbol{a} - \boldsymbol{b} = \begin{pmatrix} -1 \\ 2 \\ 5 \end{pmatrix} - \begin{pmatrix} 4 \\ -2 \\ -1 \end{pmatrix} = \begin{pmatrix} -5 \\ 4 \\ 6 \end{pmatrix}$$

so $|\boldsymbol{a} - \boldsymbol{b}| = \sqrt{(-5)^2 + 4^2 + 6^2}$

$= \sqrt{25 + 16 + 36} = \boldsymbol{\sqrt{77}}$

# Quick Test 13

**1.** $\boldsymbol{v} = \begin{pmatrix} -3 \\ 4 \\ 2 \end{pmatrix}$ and $\boldsymbol{w} = \begin{pmatrix} 1 \\ 0 \\ -5 \end{pmatrix}$

a) Write down the components of $-\boldsymbol{v} - \boldsymbol{w}$

b) Find the exact value of $|\boldsymbol{w} - \boldsymbol{v}|$

# Position vectors and applications

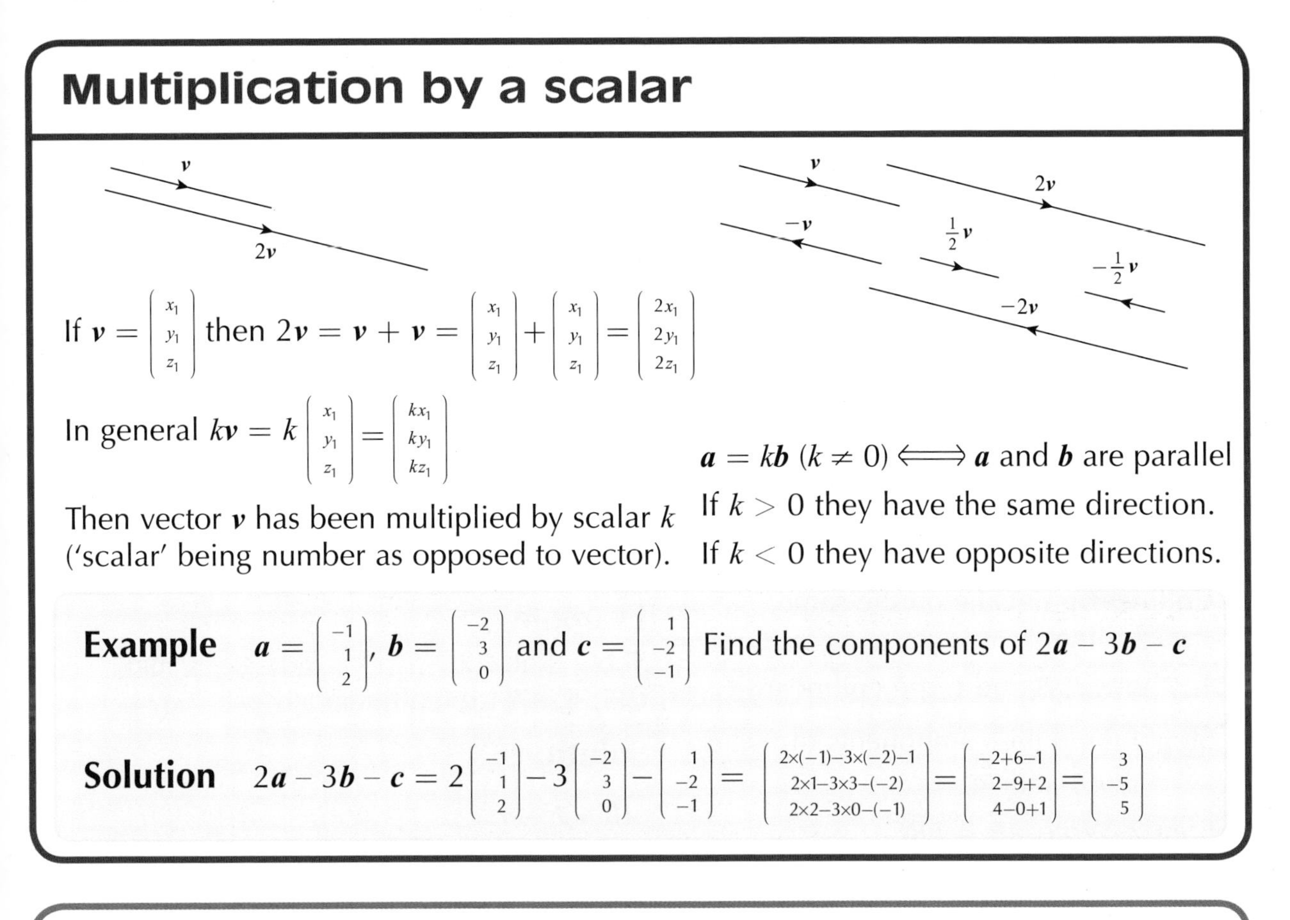


## Multiplication by a scalar

If $\boldsymbol{v} = \begin{pmatrix} x_1 \\ y_1 \\ z_1 \end{pmatrix}$ then $2\boldsymbol{v} = \boldsymbol{v} + \boldsymbol{v} = \begin{pmatrix} x_1 \\ y_1 \\ z_1 \end{pmatrix} + \begin{pmatrix} x_1 \\ y_1 \\ z_1 \end{pmatrix} = \begin{pmatrix} 2x_1 \\ 2y_1 \\ 2z_1 \end{pmatrix}$

In general $k\boldsymbol{v} = k\begin{pmatrix} x_1 \\ y_1 \\ z_1 \end{pmatrix} = \begin{pmatrix} kx_1 \\ ky_1 \\ kz_1 \end{pmatrix}$

Then vector $\boldsymbol{v}$ has been multiplied by scalar $k$ ('scalar' being number as opposed to vector).

$\boldsymbol{a} = k\boldsymbol{b}$ $(k \neq 0) \Longleftrightarrow$ $\boldsymbol{a}$ and $\boldsymbol{b}$ are parallel

If $k > 0$ they have the same direction.

If $k < 0$ they have opposite directions.

**Example** $\boldsymbol{a} = \begin{pmatrix} -1 \\ 1 \\ 2 \end{pmatrix}$, $\boldsymbol{b} = \begin{pmatrix} -2 \\ 3 \\ 0 \end{pmatrix}$ and $\boldsymbol{c} = \begin{pmatrix} 1 \\ -2 \\ -1 \end{pmatrix}$ Find the components of $2\boldsymbol{a} - 3\boldsymbol{b} - \boldsymbol{c}$

**Solution** $2\boldsymbol{a} - 3\boldsymbol{b} - \boldsymbol{c} = 2\begin{pmatrix} -1 \\ 1 \\ 2 \end{pmatrix} - 3\begin{pmatrix} -2 \\ 3 \\ 0 \end{pmatrix} - \begin{pmatrix} 1 \\ -2 \\ -1 \end{pmatrix} = \begin{pmatrix} 2\times(-1)-3\times(-2)-1 \\ 2\times1-3\times3-(-2) \\ 2\times2-3\times0-(-1) \end{pmatrix} = \begin{pmatrix} -2+6-1 \\ 2-9+2 \\ 4-0+1 \end{pmatrix} = \begin{pmatrix} 3 \\ -5 \\ 5 \end{pmatrix}$

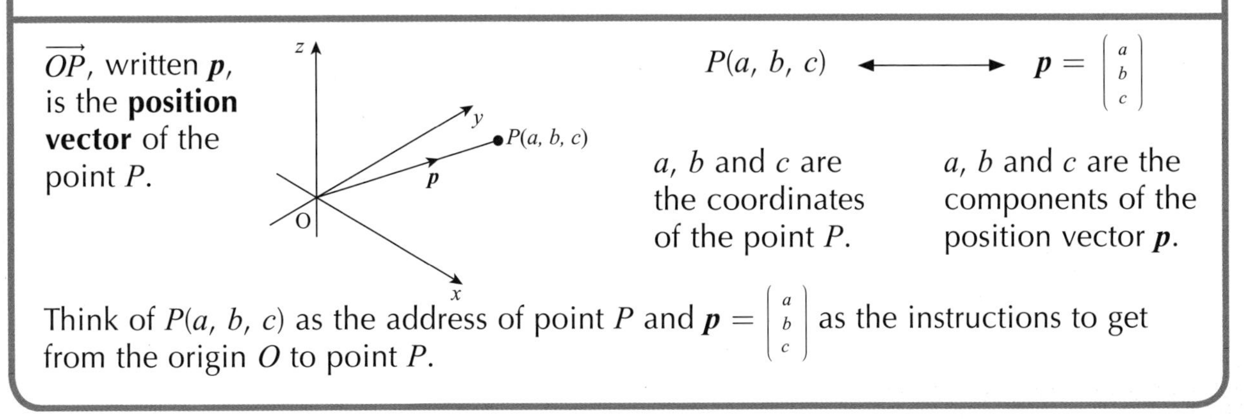


## What is a position vector?

$\overrightarrow{OP}$, written $\boldsymbol{p}$, is the **position vector** of the point $P$.

$P(a, b, c) \longleftrightarrow \boldsymbol{p} = \begin{pmatrix} a \\ b \\ c \end{pmatrix}$

$a$, $b$ and $c$ are the coordinates of the point $P$.

$a$, $b$ and $c$ are the components of the position vector $\boldsymbol{p}$.

Think of $P(a, b, c)$ as the address of point $P$ and $\boldsymbol{p} = \begin{pmatrix} a \\ b \\ c \end{pmatrix}$ as the instructions to get from the origin $O$ to point $P$.

## Using position vectors to find components

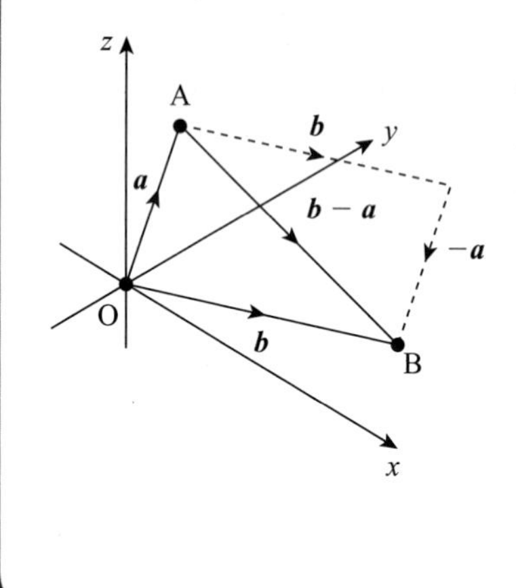


$\overrightarrow{AB} = \boldsymbol{b} - \boldsymbol{a}$

This is $\overrightarrow{AB}$ in terms of position vectors $\boldsymbol{a}$ and $\boldsymbol{b}$.

For other points:

$\overrightarrow{PQ} = \boldsymbol{q} - \boldsymbol{p}$

$\overrightarrow{RS} = \boldsymbol{s} - \boldsymbol{r}$ etc

### Example

$P$ has coordinates (1, –1, 3) and $Q$ has coordinates (–2, 5, 6). Find the components of $\overrightarrow{PQ}$.

### Solution

$\overrightarrow{PQ} = \boldsymbol{q} - \boldsymbol{p}$ (In terms of position vectors)

$$= \begin{pmatrix} -2 \\ 5 \\ 6 \end{pmatrix} - \begin{pmatrix} 1 \\ -1 \\ 3 \end{pmatrix} = \begin{pmatrix} \mathbf{-3} \\ \mathbf{6} \\ \mathbf{3} \end{pmatrix}$$

## Collinear points

**Collinear** points lie on the same straight line.

To show $A$, $B$ and $C$ are collinear:

**Step 1** Find the components of $\overrightarrow{AB}$ and $\overrightarrow{BC}$

**Step 2** Write $\overrightarrow{AB} = k\overrightarrow{BC}$ (or $\overrightarrow{BC} = k\overrightarrow{AB}$) for some constant $k$

**Step 3** State: 'Since $AB$ and $BC$ are parallel and have point $B$ in common then $A$, $B$ and $C$ are collinear'.

### Example

Show that $A(2, -1, -1)$, $B(4, 3, -5)$ and $C(5, 5, -7)$ are collinear

### Solution

$$\overrightarrow{AB} = \boldsymbol{b} - \boldsymbol{a} = \begin{pmatrix} 4 \\ 3 \\ -5 \end{pmatrix} - \begin{pmatrix} 2 \\ -1 \\ -1 \end{pmatrix} = \begin{pmatrix} 2 \\ 4 \\ -4 \end{pmatrix}$$

$$\overrightarrow{BC} = \boldsymbol{c} - \boldsymbol{b} = \begin{pmatrix} 5 \\ 5 \\ -7 \end{pmatrix} - \begin{pmatrix} 4 \\ 3 \\ -5 \end{pmatrix} = \begin{pmatrix} 1 \\ 2 \\ -2 \end{pmatrix}$$

$$\overrightarrow{AB} = 2\overrightarrow{BC}$$

Since $AB$ and $BC$ are parallel and have point $B$ in common then $A$, $B$ and $C$ are collinear

For example $\overrightarrow{AB} = 2\overrightarrow{BC}$

A ——— 2 units ——— B ——— 1 unit ——— C

Note: In this case $B$ divides $AC$ in the ratio 2 : 1

For $\overrightarrow{AB} = k\overrightarrow{BC}$ the ratio is $k : 1$

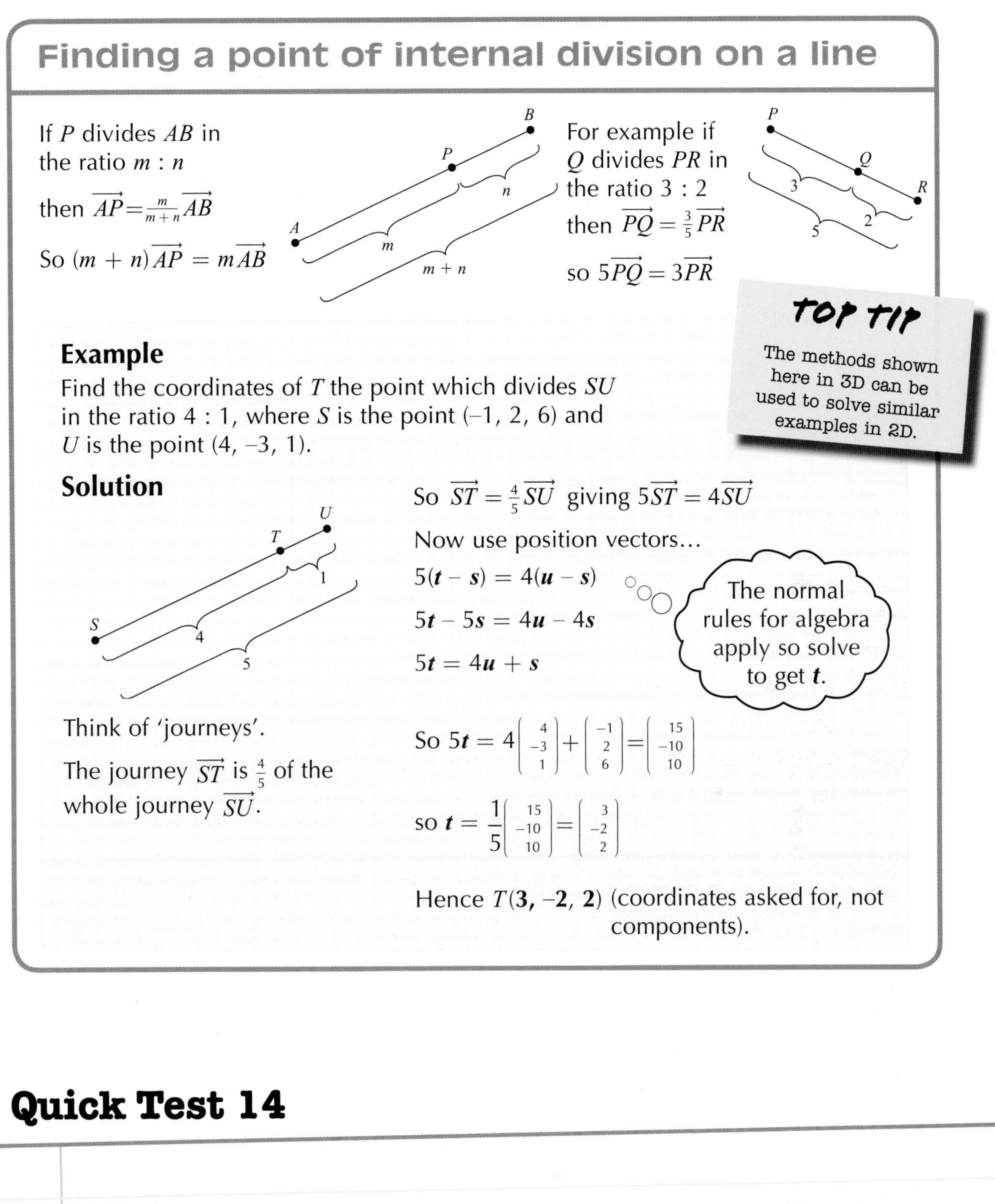

## Finding a point of internal division on a line

If $P$ divides $AB$ in the ratio $m : n$

then $\overrightarrow{AP} = \frac{m}{m+n}\overrightarrow{AB}$

So $(m+n)\overrightarrow{AP} = m\overrightarrow{AB}$



For example if $Q$ divides $PR$ in the ratio 3 : 2

then $\overrightarrow{PQ} = \frac{3}{5}\overrightarrow{PR}$

so $5\overrightarrow{PQ} = 3\overrightarrow{PR}$



**TOP TIP**

The methods shown here in 3D can be used to solve similar examples in 2D.

**Example**

Find the coordinates of $T$ the point which divides $SU$ in the ratio 4 : 1, where $S$ is the point (–1, 2, 6) and $U$ is the point (4, –3, 1).

**Solution**



Think of 'journeys'.

The journey $\overrightarrow{ST}$ is $\frac{4}{5}$ of the whole journey $\overrightarrow{SU}$.

So $\overrightarrow{ST} = \frac{4}{5}\overrightarrow{SU}$ giving $5\overrightarrow{ST} = 4\overrightarrow{SU}$

Now use position vectors…

$5(\mathbf{t} - \mathbf{s}) = 4(\mathbf{u} - \mathbf{s})$

$5\mathbf{t} - 5\mathbf{s} = 4\mathbf{u} - 4\mathbf{s}$

$5\mathbf{t} = 4\mathbf{u} + \mathbf{s}$

The normal rules for algebra apply so solve to get $\mathbf{t}$.

So $5\mathbf{t} = 4\begin{pmatrix}4\\-3\\1\end{pmatrix} + \begin{pmatrix}-1\\2\\6\end{pmatrix} = \begin{pmatrix}15\\-10\\10\end{pmatrix}$

so $\mathbf{t} = \frac{1}{5}\begin{pmatrix}15\\-10\\10\end{pmatrix} = \begin{pmatrix}3\\-2\\2\end{pmatrix}$

Hence $T$(**3, –2, 2**) (coordinates asked for, not components).

# Quick Test 14

1. The points $P(-1, 0, 2)$, $Q(3, 5, -2)$, $R(-1, 6, -7)$ and $S(-5, 1, -3)$ form a quadrilateral.
   a) Find the components of: (i) $\overrightarrow{PQ}$ (ii) $\overrightarrow{SR}$
   b) What sort of quadrilateral is $PQRS$?
2. Show that $A(2, -1, 4)$, $B(3, 2, 5)$ and $C(6, 11, 8)$ are collinear and find the ratio in which the 'middle' point divides the line joining the 'end' points (in the given order).

# The scalar product and applications

## The angle between two vectors

Place the two vectors tail-to-tail:

$\theta$ is the angle between $\boldsymbol{a}$ and $\boldsymbol{b}$

$\theta$ always lies in the range $0 \leq \theta \leq \pi$

Maximum angle is $\pi$

Minimum angle is 0

## The scalar (dot) product

Using magnitudes and the angle...

Using components... $\boldsymbol{a} = \begin{pmatrix} x_1 \\ y_1 \\ z_1 \end{pmatrix}$, $\boldsymbol{b} = \begin{pmatrix} x_2 \\ y_2 \\ z_2 \end{pmatrix}$

$\boldsymbol{a}.\boldsymbol{b} = |\boldsymbol{a}|\,|\boldsymbol{b}| \cos\theta$

$\boldsymbol{a}.\boldsymbol{b} = x_1x_2 + y_1y_2 + z_1z_2$

These two calculations yield the same **number** which is written $\boldsymbol{a}.\boldsymbol{b}$ and is called the **Scalar Product** or **Dot Product** of vectors $\boldsymbol{a}$ and $\boldsymbol{b}$.

## Calculating the angle

Rearrange $\boldsymbol{a}.\boldsymbol{b} = |\boldsymbol{a}||\boldsymbol{b}| \cos\theta$ to get $\cos\theta = \dfrac{\boldsymbol{a}.\boldsymbol{b}}{|\boldsymbol{a}|\,|\boldsymbol{b}|}$

If you know the components $\boldsymbol{a} = \begin{pmatrix} x_1 \\ y_1 \\ z_1 \end{pmatrix}$ $\boldsymbol{b} = \begin{pmatrix} x_2 \\ y_2 \\ z_2 \end{pmatrix}$

then $\cos\theta = \dfrac{\boldsymbol{a}.\boldsymbol{b}}{|\boldsymbol{a}|\,|\boldsymbol{b}|} = \dfrac{x_1x_2 + y_1y_2 + z_1z_2}{\sqrt{x_1^2 + y_1^2 + z_1^2}\sqrt{x_2^2 + y_2^2 + z_2^2}}$

Remember that $\theta$ will be a 1st or 2nd quadrant angle since $0 \leq \theta \leq \pi$

**Example**

Calculate the angle $\theta$ between $\boldsymbol{a} = \begin{pmatrix} 2 \\ 3 \\ -1 \end{pmatrix}$ and $\boldsymbol{b} = \begin{pmatrix} 1 \\ -2 \\ 3 \end{pmatrix}$

**Solution**

$$|\boldsymbol{a}| = \sqrt{2^2 + 3^2 + (-1)^2} = \sqrt{14}$$

$$|\boldsymbol{b}| = \sqrt{1^2 + (-2)^2 + 3^2} = \sqrt{14}$$

$$\boldsymbol{a.b} = \begin{pmatrix} 2 \\ 3 \\ -1 \end{pmatrix} \cdot \begin{pmatrix} 1 \\ -2 \\ 3 \end{pmatrix} = 2\times1 + 3\times(-2) + (-1)\times3 = -7$$

$$\text{so } \cos\theta = \frac{\boldsymbol{a.b}}{|\boldsymbol{a}||\boldsymbol{b}|} = \frac{-7}{\sqrt{14}\sqrt{14}} = -\frac{7}{14} = -\frac{1}{2}$$

$$\text{so} \quad \theta = \pi - \frac{\pi}{3} = \frac{\mathbf{2\pi}}{\mathbf{3}} \quad \text{(second quadrant only)}$$

## Perpendicular vectors

If $\boldsymbol{a}$ and $\boldsymbol{b}$ are perpendicular then $\boldsymbol{a.b} = 0$

(since $\cos \frac{\pi}{2} = 0$)

If $\boldsymbol{a.b} = 0$ and both $\boldsymbol{a}$ and $\boldsymbol{b}$ are non-zero then $\boldsymbol{a}$ is perpendicular to $\boldsymbol{b}$

**TOP TIP**

To prove two vectors are perpendicular you show their dot product is zero

## Solving perpendicularity problems

**Example 1** A triangle $ABC$ has vertices $A(-1, 3, 2)$, $B(1, -3, 3)$ and $C(0, 2, -6)$. Show that it is right-angled.

**Solution**

$$\overrightarrow{AB} = \boldsymbol{b} - \boldsymbol{a} = \begin{pmatrix} 1 \\ -3 \\ 3 \end{pmatrix} - \begin{pmatrix} -1 \\ 3 \\ 2 \end{pmatrix} = \begin{pmatrix} 2 \\ -6 \\ 1 \end{pmatrix}$$

$$\overrightarrow{AC} = \boldsymbol{c} - \boldsymbol{a} = \begin{pmatrix} 0 \\ 2 \\ -6 \end{pmatrix} - \begin{pmatrix} -1 \\ 3 \\ 2 \end{pmatrix} = \begin{pmatrix} 1 \\ -1 \\ -8 \end{pmatrix}$$

$$\overrightarrow{AB}.\overrightarrow{AC} = \begin{pmatrix} 2 \\ -6 \\ 1 \end{pmatrix} \cdot \begin{pmatrix} 1 \\ -1 \\ -8 \end{pmatrix} = 2\times1 + (-6)\times(-1) + 1\times(-8) = 0$$

hence $\overrightarrow{AB}$ is perpendicular to $\overrightarrow{AC}$

$\angle BAC = 90°$

So $\Delta ABC$ is right-angled at $A$.

### Example 2

If $\boldsymbol{v} = \begin{pmatrix} x \\ -2 \\ 9 \end{pmatrix}$ and $\boldsymbol{w} = \begin{pmatrix} x \\ 3x \\ 1 \end{pmatrix}$ are perpendicular calculate the value of $x$

### Solution

$$\boldsymbol{v}.\boldsymbol{w} = 0 \Rightarrow \begin{pmatrix} x \\ -2 \\ 9 \end{pmatrix} \cdot \begin{pmatrix} x \\ 3x \\ 1 \end{pmatrix} = 0 \Rightarrow x \times x + (-2) \times 3x + 9 \times 1 = 0$$

$$\Rightarrow x^2 - 6x + 9 = 0 \Rightarrow (x-3)^2 = 0 \Rightarrow x = 3$$

**TOP TIP**

In the exam if you are given a vector written with unit vectors $\boldsymbol{i}, \boldsymbol{j}$ and $\boldsymbol{k}$ rewrite it in the usual component form.

## Unit vectors

A **Unit Vector** has a magnitude of 1 unit. The three unit vectors parallel to the three axes are:

$$\boldsymbol{i} = \begin{pmatrix} 1 \\ 0 \\ 0 \end{pmatrix} \quad \boldsymbol{j} = \begin{pmatrix} 0 \\ 1 \\ 0 \end{pmatrix} \quad \boldsymbol{k} = \begin{pmatrix} 0 \\ 0 \\ 1 \end{pmatrix}$$

z
k
y
j
O
i
x

These form a set of **basis vectors** since any vector $\boldsymbol{v}$ can be written in terms of them.

If $\boldsymbol{v} = \begin{pmatrix} a \\ b \\ c \end{pmatrix} = a\begin{pmatrix} 1 \\ 0 \\ 0 \end{pmatrix} + b\begin{pmatrix} 0 \\ 1 \\ 0 \end{pmatrix} + c\begin{pmatrix} 0 \\ 0 \\ 1 \end{pmatrix}$

then $\boldsymbol{v} = a\boldsymbol{i} + b\boldsymbol{j} + c\boldsymbol{k}$

### Example

$\boldsymbol{v} = \boldsymbol{i} - 3\boldsymbol{k}$ and $\boldsymbol{w} = 5\boldsymbol{i} - 2\boldsymbol{j} + \boldsymbol{k}$. Find a **unit** vector parallel to vector $\boldsymbol{v} - \boldsymbol{w}$.

### Solution

$$\boldsymbol{v} - \boldsymbol{w} = \begin{pmatrix} 1 \\ 0 \\ -3 \end{pmatrix} - \begin{pmatrix} 5 \\ -2 \\ 1 \end{pmatrix} = \begin{pmatrix} -4 \\ 2 \\ -4 \end{pmatrix}$$

so $|\boldsymbol{v} - \boldsymbol{w}| = \sqrt{(-4)^2 + 2^2 + (-4)^2}$

$= \sqrt{36} = 6$

so $\frac{1}{6}(\boldsymbol{v} - \boldsymbol{w})$ has magnitude 1 unit.

$$\frac{1}{6}\begin{pmatrix} -4 \\ 2 \\ -4 \end{pmatrix} = \begin{pmatrix} -\frac{2}{3} \\ \frac{1}{3} \\ -\frac{2}{3} \end{pmatrix} = -\frac{2}{3}i + \frac{1}{3}j - \frac{2}{3}k$$

is the required unit vector.

# Quick Test 15

1. Find the size of angle KLM where K(–2, 5, 4), L(–1, 0, –3) and M(2, 2, 8).
2. The vectors $\begin{pmatrix} m \\ -2 \\ -1 \end{pmatrix}$ and $\begin{pmatrix} m-1 \\ 1 \\ 4 \end{pmatrix}$ are perpendicular. Find the two possible values for $m$.
3. Find a unit vector parallel to $\boldsymbol{u} + \boldsymbol{v}$ where: $\boldsymbol{u} = 3\boldsymbol{i} - \boldsymbol{j} + 2\boldsymbol{k}$ and $\boldsymbol{v} = -4\boldsymbol{i} - \boldsymbol{j}$

# Working with vectors

## Algebraic properties

Most of the 'normal' rules of algebra apply to vectors.

For instance:

$2(\boldsymbol{b} - \boldsymbol{a}) = 3(\boldsymbol{c} - \boldsymbol{a})$ rearranges to $\boldsymbol{a} = 3\boldsymbol{c} - 2\boldsymbol{b}$

$-5(\boldsymbol{v} - \boldsymbol{w}) = -5\boldsymbol{v} + 5\boldsymbol{w}$,

$-(\boldsymbol{a} + \boldsymbol{b}) = -\boldsymbol{a} - \boldsymbol{b}$ etc

For any three vectors $\boldsymbol{a}$, $\boldsymbol{b}$ and $\boldsymbol{c}$ that are non-zero

$$\boldsymbol{a}.(\boldsymbol{b} + \boldsymbol{c}) = \boldsymbol{a}.\boldsymbol{b} + \boldsymbol{a}.\boldsymbol{c}$$

Also

$$\boldsymbol{a}.\boldsymbol{a} = |\boldsymbol{a}||\boldsymbol{a}| \cos 0 = |\boldsymbol{a}||\boldsymbol{a}| \times 1 = |\boldsymbol{a}|^2$$

and

$$\boldsymbol{a}.\boldsymbol{b} = \boldsymbol{b}.\boldsymbol{a}$$

***Warning***

$\boldsymbol{a}^2$, $(\boldsymbol{a} + \boldsymbol{b})^2$, $\sqrt{\boldsymbol{a}}$ are all meaningless.

For the basis vectors $\boldsymbol{i}$, $\boldsymbol{j}$ and $\boldsymbol{k}$ you get:

$$\boldsymbol{i}.\boldsymbol{i} = \boldsymbol{j}.\boldsymbol{j} = \boldsymbol{k}.\boldsymbol{k} = 1 \times 1 \times \cos 0 = 1 \times 1 \times 1 = 1$$

$$\boldsymbol{i}.\boldsymbol{j} = \boldsymbol{j}.\boldsymbol{k} = \boldsymbol{i}.\boldsymbol{k} = 1 \times 1 \times \cos \frac{\pi}{2} = 1 \times 1 \times 0 = 0$$

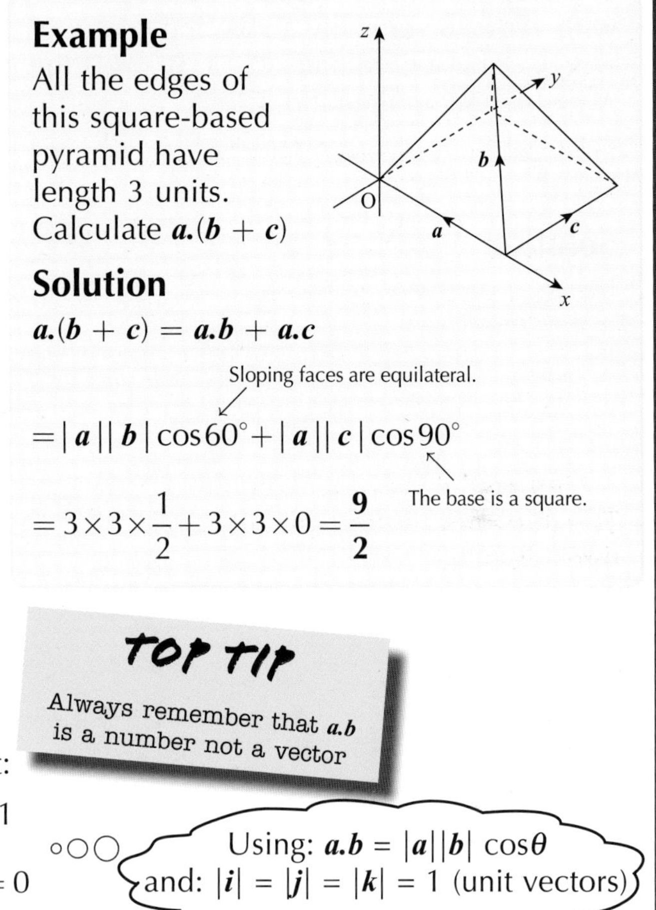


### Example

All the edges of this square-based pyramid have length 3 units. Calculate $\boldsymbol{a}.(\boldsymbol{b} + \boldsymbol{c})$

### Solution

$$\boldsymbol{a}.(\boldsymbol{b} + \boldsymbol{c}) = \boldsymbol{a}.\boldsymbol{b} + \boldsymbol{a}.\boldsymbol{c}$$

Sloping faces are equilateral.

$$= |\boldsymbol{a}||\boldsymbol{b}| \cos 60° + |\boldsymbol{a}||\boldsymbol{c}| \cos 90°$$

The base is a square.

$$= 3 \times 3 \times \frac{1}{2} + 3 \times 3 \times 0 = \frac{\mathbf{9}}{\mathbf{2}}$$

**TOP TIP**

Always remember that $\boldsymbol{a}.\boldsymbol{b}$ is a number not a vector

Using: $\boldsymbol{a}.\boldsymbol{b} = |\boldsymbol{a}||\boldsymbol{b}| \cos\theta$
and: $|\boldsymbol{i}| = |\boldsymbol{j}| = |\boldsymbol{k}| = 1$ (unit vectors)

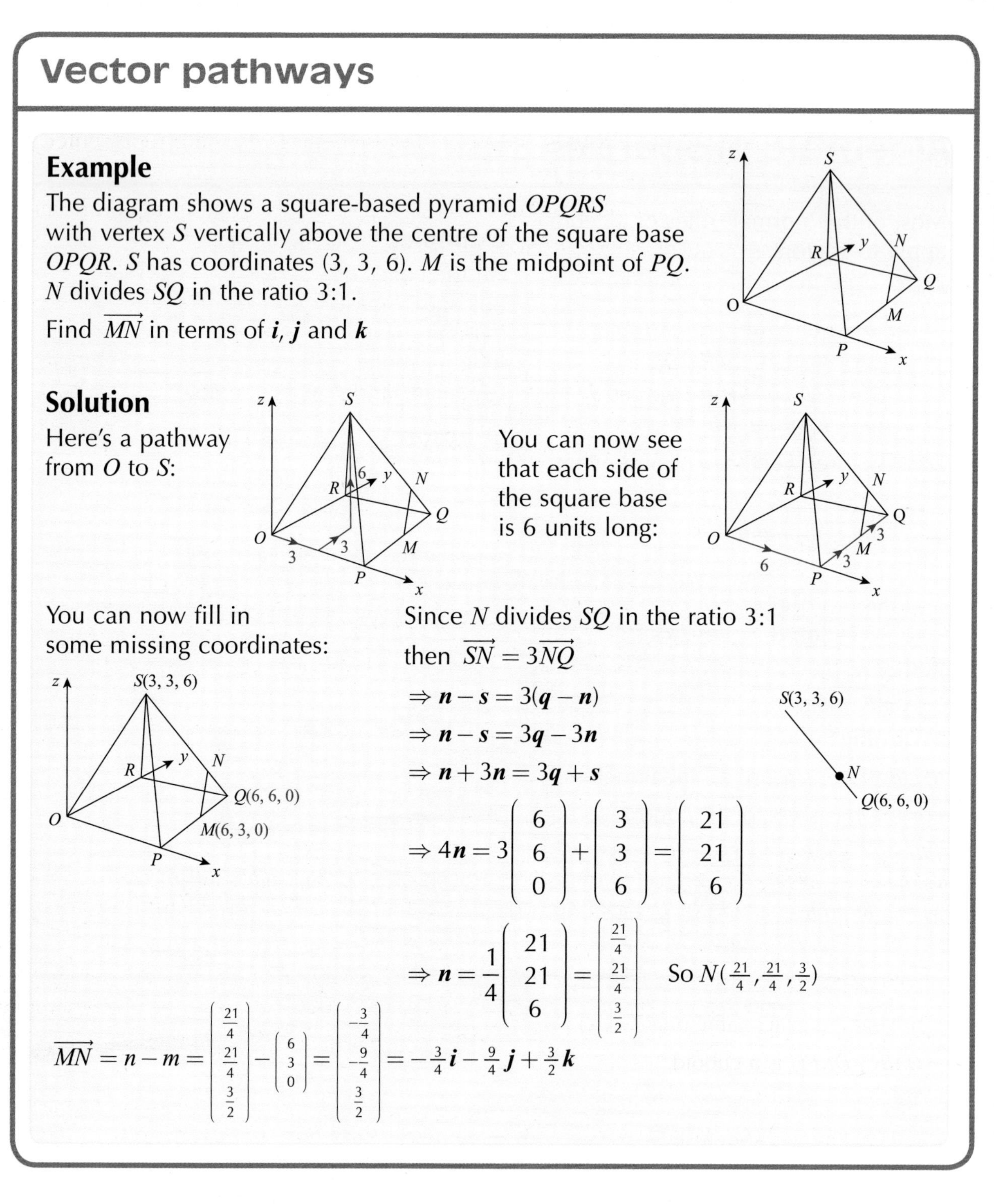

## Vector pathways

### Example

The diagram shows a square-based pyramid $OPQRS$ with vertex $S$ vertically above the centre of the square base $OPQR$. $S$ has coordinates (3, 3, 6). $M$ is the midpoint of $PQ$. $N$ divides $SQ$ in the ratio 3:1.

Find $\overrightarrow{MN}$ in terms of $\boldsymbol{i}$, $\boldsymbol{j}$ and $\boldsymbol{k}$



### Solution

Here's a pathway from $O$ to $S$:



You can now see that each side of the square base is 6 units long:



You can now fill in some missing coordinates:



Since $N$ divides $SQ$ in the ratio 3:1

then $\overrightarrow{SN} = 3\overrightarrow{NQ}$

$\Rightarrow \boldsymbol{n} - \boldsymbol{s} = 3(\boldsymbol{q} - \boldsymbol{n})$

$\Rightarrow \boldsymbol{n} - \boldsymbol{s} = 3\boldsymbol{q} - 3\boldsymbol{n}$

$\Rightarrow \boldsymbol{n} + 3\boldsymbol{n} = 3\boldsymbol{q} + \boldsymbol{s}$



$$\Rightarrow 4\boldsymbol{n} = 3\begin{pmatrix} 6 \\ 6 \\ 0 \end{pmatrix} + \begin{pmatrix} 3 \\ 3 \\ 6 \end{pmatrix} = \begin{pmatrix} 21 \\ 21 \\ 6 \end{pmatrix}$$

$$\Rightarrow \boldsymbol{n} = \frac{1}{4}\begin{pmatrix} 21 \\ 21 \\ 6 \end{pmatrix} = \begin{pmatrix} \frac{21}{4} \\ \frac{21}{4} \\ \frac{3}{2} \end{pmatrix}$$

So $N(\frac{21}{4}, \frac{21}{4}, \frac{3}{2})$

$$\overrightarrow{MN} = \boldsymbol{n} - \boldsymbol{m} = \begin{pmatrix} \frac{21}{4} \\ \frac{21}{4} \\ \frac{3}{2} \end{pmatrix} - \begin{pmatrix} 6 \\ 3 \\ 0 \end{pmatrix} = \begin{pmatrix} -\frac{3}{4} \\ \frac{9}{4} \\ \frac{3}{2} \end{pmatrix} = -\frac{3}{4}\boldsymbol{i} - \frac{9}{4}\boldsymbol{j} + \frac{3}{2}\boldsymbol{k}$$

## Forces in equilibrium

There are three forces acting on an object at $A$. They are represented by vectors $\boldsymbol{f}_1$, $\boldsymbol{f}_2$ and $\boldsymbol{f}_3$.

If the object is stationary then the system is said to be in equilibrium.

In this case the vector sum of the forces is zero:

$$\boldsymbol{f}_1 + \boldsymbol{f}_2 + \boldsymbol{f}_3 = \boldsymbol{0}$$

**Example**

Three forces $\boldsymbol{f}_p = \begin{pmatrix} 3 \\ y \\ -1 \end{pmatrix}$, $\boldsymbol{f}_q = \begin{pmatrix} 3 \\ -2 \\ z \end{pmatrix}$ and $\boldsymbol{f}_r = \begin{pmatrix} x \\ 4 \\ -1 \end{pmatrix}$ acting on an object are in equilibrium. Find the values of $x$, $y$ and $z$.

**Solution**

Since the system is in equilibrium $\boldsymbol{f}_p + \boldsymbol{f}_q + \boldsymbol{f}_r = \boldsymbol{0}$

So $\begin{pmatrix} 3 \\ y \\ -1 \end{pmatrix} + \begin{pmatrix} 3 \\ -2 \\ z \end{pmatrix} + \begin{pmatrix} x \\ 4 \\ -1 \end{pmatrix} = \begin{pmatrix} 0 \\ 0 \\ 0 \end{pmatrix} \Rightarrow \begin{cases} 3+3+x=0 \Rightarrow 6+x=0 \Rightarrow x=-6 \\ y-2+4=0 \Rightarrow y+2=0 \Rightarrow y=-2 \\ -1+z-1=0 \Rightarrow z-2=0 \Rightarrow z=2 \end{cases}$

# Quick Test 16

**Top Tip**

It is difficult to write in bold font. When you write down a vector $\boldsymbol{v}$ you should underline it $\underline{v}$

1. Each edge of this cube has length 1 unit.

   a) Find the exact value of $|\boldsymbol{b}|$ and $|\boldsymbol{c}|$

   b) Find the exact value of $\boldsymbol{a}.(\boldsymbol{b} + \boldsymbol{c})$

2. $OABC$, $DEFG$ is a cuboid.

   The vertex $F$ is the point $(5, 6, 2)$.

   $M$ is the midpoint of $DG$.

   $N$ divides $AB$ in the ratio 1:2.

   a) Find the coordinates of $M$ and $N$.

   b) Write $\overrightarrow{MN}$ in terms of unit vectors $\boldsymbol{i}$, $\boldsymbol{j}$ and $\boldsymbol{k}$

3. Four forces act on an object and are in equilibrium.

   Three of the forces are: $\boldsymbol{F}_1 = \begin{pmatrix} -2 \\ 5 \\ 0 \end{pmatrix}$, $\boldsymbol{F}_2 = \begin{pmatrix} 3 \\ -3 \\ -1 \end{pmatrix}$ and $\boldsymbol{F}_3 = \begin{pmatrix} 2 \\ 1 \\ -1 \end{pmatrix}$

   a) Find the components of the fourth force $\boldsymbol{F}_4$

   b) Find the magnitude of $\boldsymbol{F}_4$

# Sample Unit 1 test questions

## 1.1 Applying algebraic skills to logarithms and exponentials

1. Simplify $\log_2 3ab - \log_2 3b$
2. Express $\log_b a^3 + \log_b a^2$ in the form $m \log_b a$
3. Solve $\log_2 (y - 3) = 1$

## 1.2 Applying trig skills to manipulating expressions

1. Express $\sin x° + 3\cos x°$ in the form $k\cos(x - a)°$ where $k > 0$ and $0 \leq a < 360$
2. Show that $(1 - \sin x)^2 + 2\sin x \equiv 2 - \cos^2 x$
3. Use the information in the diagram to find the exact value of $\sin(a + b)$

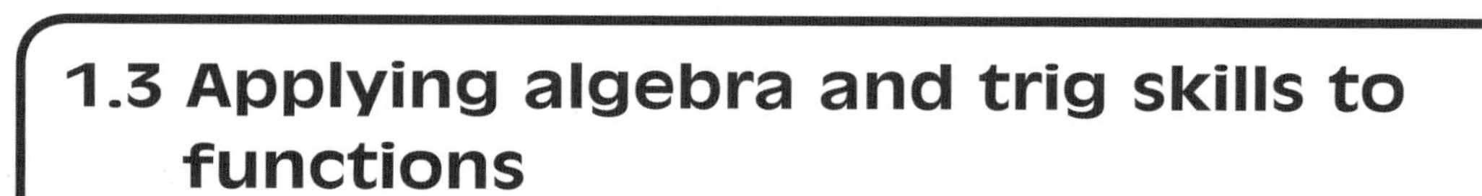

## 1.3 Applying algebra and trig skills to functions

1. Sketch the graph $y = k\cos(x + \frac{\pi}{4})$ for $0 \leq x < 2\pi$ and $k > 0$.

   Show the minimum and maximum values and where the graph cuts the $x$-axis.
2. The graph of $y = f(x)$ is shown.
   It passes through the points $A(0, 2)$, $B(1, 0)$, $C(2, 1)$ and $D(3, 0)$.

   Sketch the graph $y = f(x + 1) - 2$

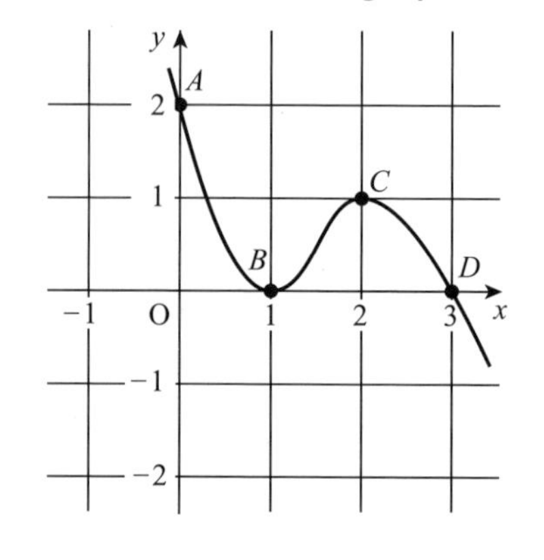


3. Find $f^{-1}(x)$ where $f(x) = \frac{1}{2}x + 3$

4. The diagram shows part of the graph of $y = k\cos ax + b$. Write down the values of $k$, $a$ and $b$.

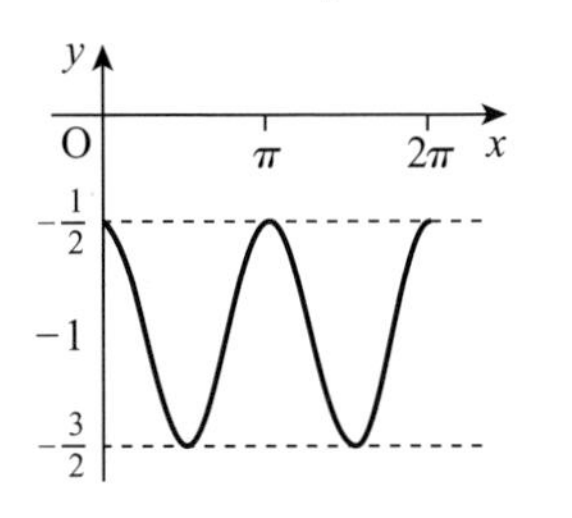


5. The graph $y = \log_m (x + n)$ is shown. Find the values of $m$ and $n$.

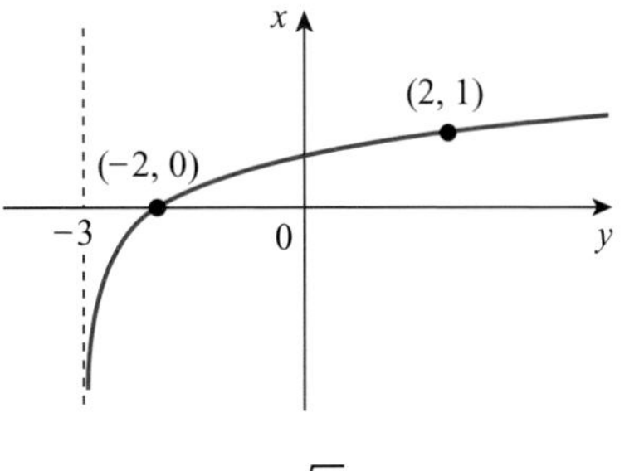


6. Functions $f$ and $g$ are given by $f(x) = 2 - x$ and $g(x) = -\sqrt{x}$
   a) Find an expression for $f(g(x))$
   b) State a suitable domain for $f(g(x))$

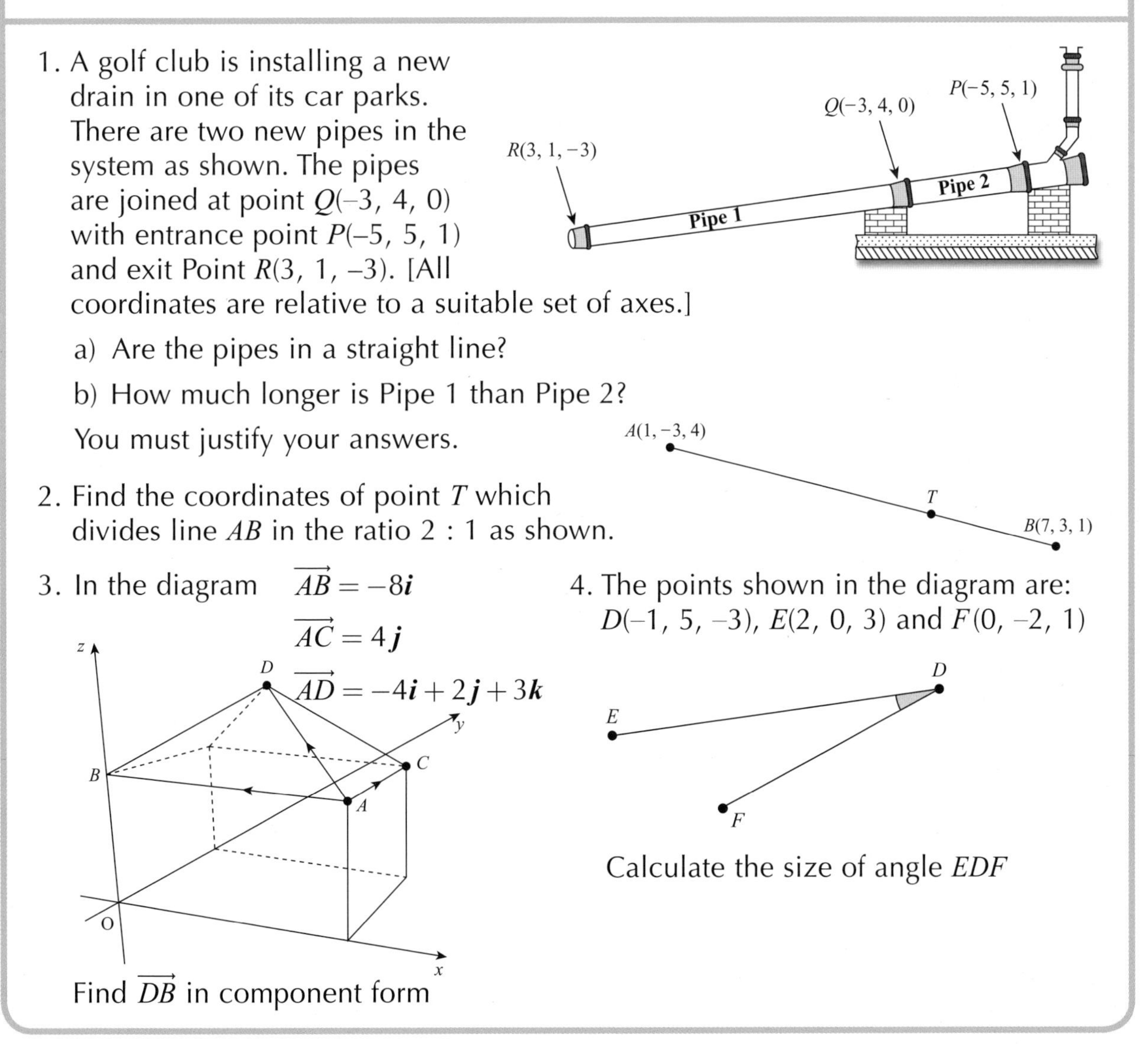

## 1.4 Applying geometric skills to vectors

1. A golf club is installing a new drain in one of its car parks. There are two new pipes in the system as shown. The pipes are joined at point $Q(-3, 4, 0)$ with entrance point $P(-5, 5, 1)$ and exit Point $R(3, 1, -3)$. [All coordinates are relative to a suitable set of axes.]



   a) Are the pipes in a straight line?
   b) How much longer is Pipe 1 than Pipe 2?
   You must justify your answers.

2. Find the coordinates of point $T$ which divides line $AB$ in the ratio 2 : 1 as shown.



3. In the diagram $\overrightarrow{AB} = -8\boldsymbol{i}$

$\overrightarrow{AC} = 4\boldsymbol{j}$

$\overrightarrow{AD} = -4\boldsymbol{i} + 2\boldsymbol{j} + 3\boldsymbol{k}$



Find $\overrightarrow{DB}$ in component form

4. The points shown in the diagram are: $D(-1, 5, -3)$, $E(2, 0, 3)$ and $F(0, -2, 1)$



Calculate the size of angle $EDF$

# Sample end-of-course exam questions (Unit 1)

## Non-calculator

1. $g(x) = 3x$ and $h(x) = \sin 2x$ are two functions defined on suitable domains.
   What is the value of $h\left(g\left(\frac{\pi}{12}\right)\right)$?

2. If the exact value of $\tan x$ is $\frac{1}{\sqrt{3}}$, where $0 \leq x \leq \frac{\pi}{2}$, find the exact value of $\sin 2x$.

3. $OABC$ is a tetrahedron. $A$ is the point $(4, -2, 0)$, $B$ is $(3, 5, 0)$ and $C$ is $(1, 4, 6)$. $P$ divides $AC$ in the ratio 2:1.

   a) Find the coordinates of $P$.

   b) Express $\overrightarrow{BP}$ in terms of the unit vectors $\boldsymbol{i}$, $\boldsymbol{j}$ and $\boldsymbol{k}$.

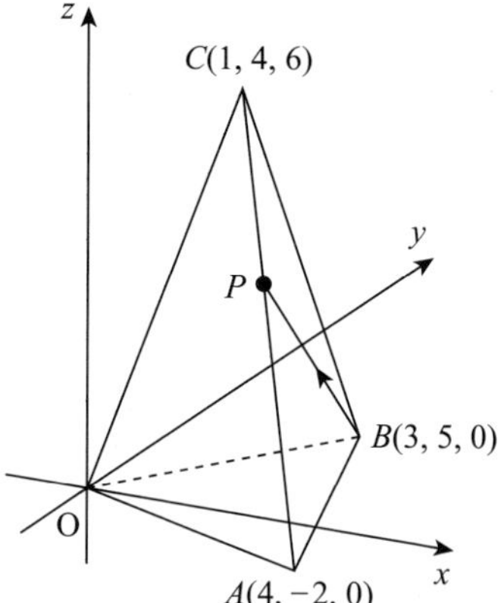


4. Express $2(\sqrt{3}\cos x - \sin x)$ in the form $k\cos(x + a)$ where $k > 0$ and $0 \leq a < 2\pi$.

5. $ABC$ is a right-angled triangle as shown in the diagram, with vertices $A(1, -2, -k)$, $B(k, k, 0)$ and $C(4, -3, 3 - k)$.

   a) Find the value of $k$.

   $D$ is the point $(13, -6, 11)$.

   b) Show that $A$, $C$ and $D$ are collinear and find the ratio in which $C$ divides $AD$.

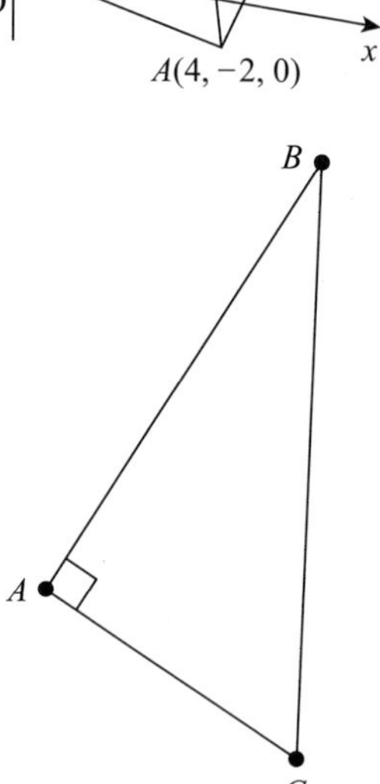


6. The diagram shows the graph with equation $y = 2\sin ax + b$ where $a$ and $b$ are constants and $0 \leq x \leq \pi$.

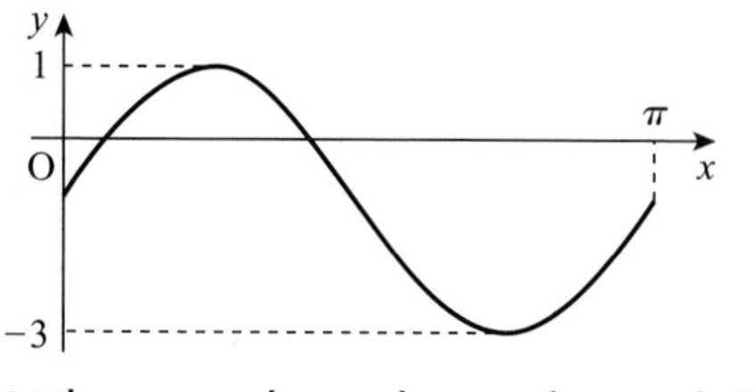


   What are the values of $a$ and $b$?

7. a) If $\log_{\sqrt{a}} b = 2c$ show that $\log_a b = c$.

   b) Hence show that $\log_5 7 - \log_{25} 7 = \log_{25} 7$.

## Calculator allowed

1. a) Simplify $\frac{2x^2-7x+6}{x^2-4}$.

   b) Solve $\log_3(2x^2-7x+6)-\log_3(x^2-4)=2$.

2. The diagram shows a cuboid surmounted by a pyramid. One of the triangular faces of the pyramid has vertices $A(3, 2, 5)$, $B(6, 0, 3)$ and $C(6, 4, 3)$.

   (a) Express $\overrightarrow{CA}$ and $\overrightarrow{CB}$ in component form.

   (b) Calculate the angle between the two edges $CA$ and $CB$.

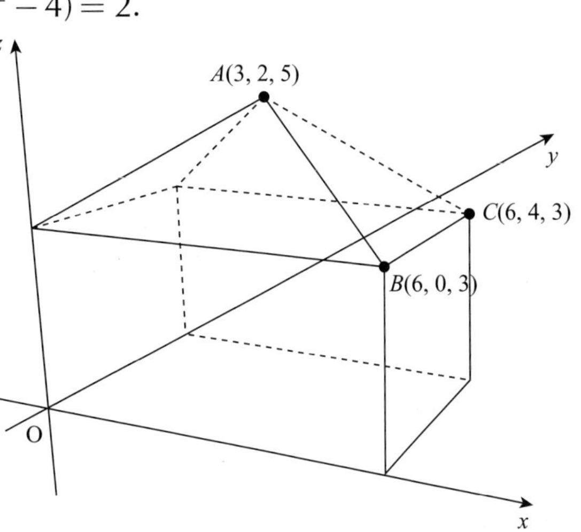


3. The formula $P=6e^{0.0138t}$ is used to predict the population $P$ of the world, in billions, $t$ years after January 1, 2000.

   (a) What was the population of the world on January 1, 2000?

   (b) At the start of which year will the world's population be more than double that of the population on January 1, 2000?

4. Two right-angled triangles $PQR$ and $RPS$ have lengths as shown in the diagram.

   Angle $PRQ = a°$ and angle $PRS = b°$.

   a) Show that the exact value of $\cos(a+b)°$ is $-\frac{33}{65}$.

   b) Calculate the exact value of $\sin(a+b)°$.

   c) Hence calculate the exact value of $\tan(a+b)°$.

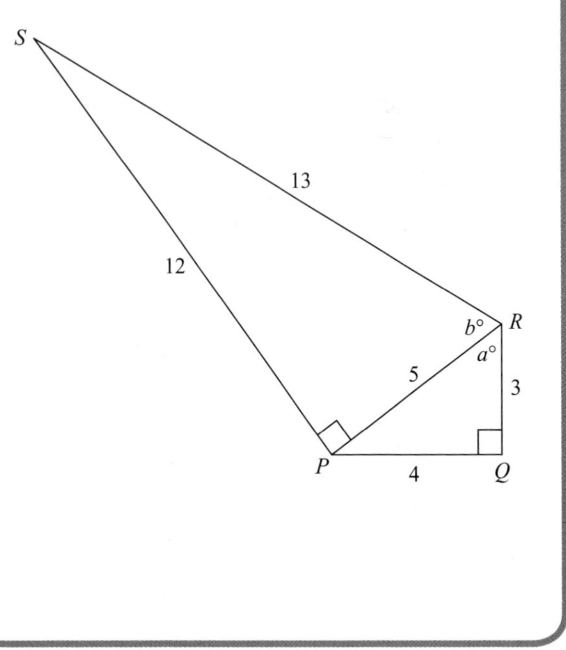


5. The functions $f$ and $g$ are defined on a suitable domain by $f(x)=\frac{2}{x+1}$ and $g(x)=2x+1$.

   Prove that $f(g(x))=\frac{1}{2}f(x)$.

# Quadratic theory revisited I

## Roots of quadratic equations

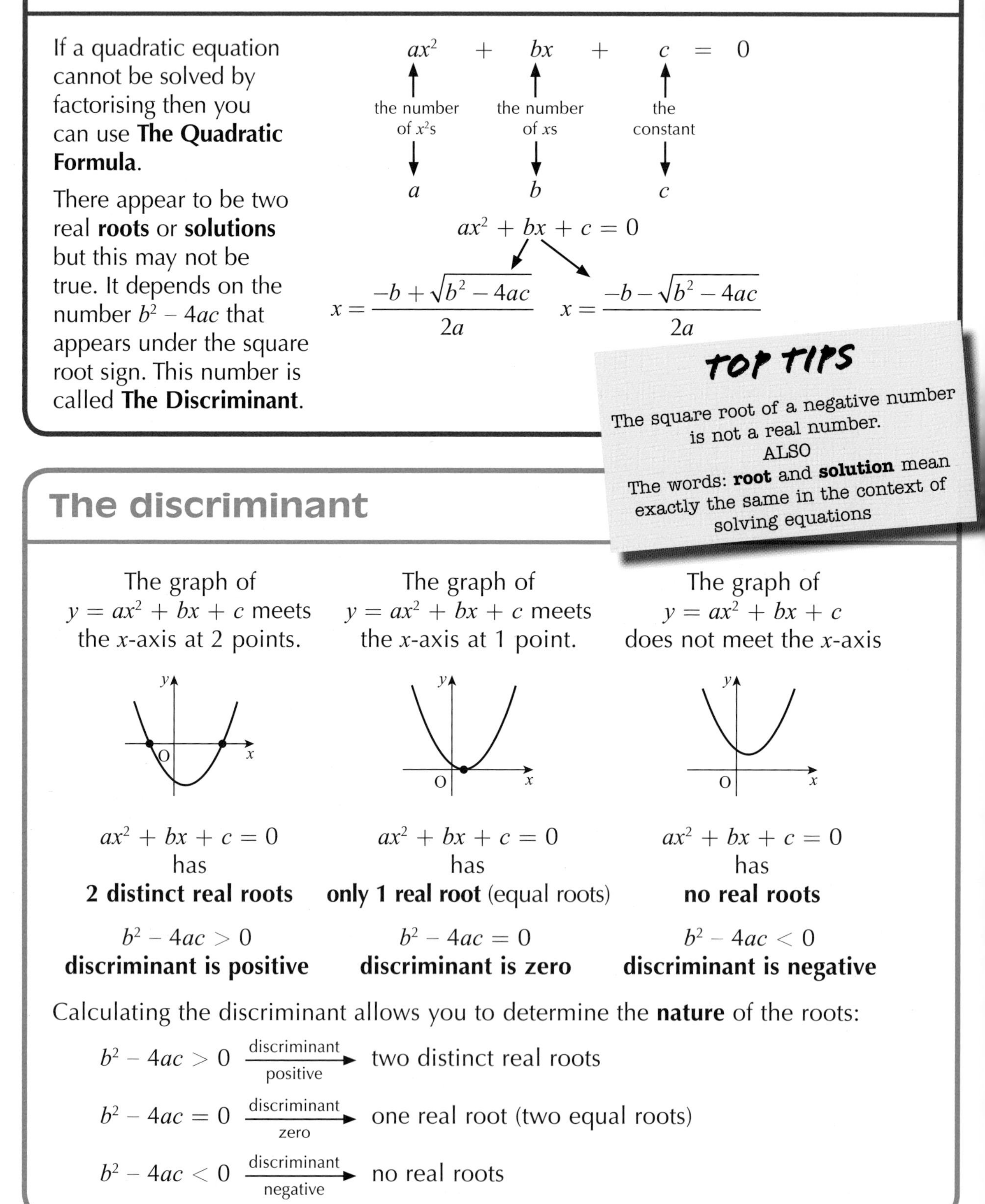

If a quadratic equation cannot be solved by factorising then you can use **The Quadratic Formula**.

There appear to be two real **roots** or **solutions** but this may not be true. It depends on the number $b^2 - 4ac$ that appears under the square root sign. This number is called **The Discriminant**.

$$ax^2 + bx + c = 0$$

- $ax^2$: the number of $x^2$s → $a$
- $bx$: the number of $x$s → $b$
- $c$: the constant → $c$

$$ax^2 + bx + c = 0$$

$$x = \frac{-b + \sqrt{b^2 - 4ac}}{2a} \qquad x = \frac{-b - \sqrt{b^2 - 4ac}}{2a}$$

**TOP TIPS**

The square root of a negative number is not a real number.
ALSO
The words: **root** and **solution** mean exactly the same in the context of solving equations

## The discriminant

| The graph of $y = ax^2 + bx + c$ meets the $x$-axis at 2 points. | The graph of $y = ax^2 + bx + c$ meets the $x$-axis at 1 point. | The graph of $y = ax^2 + bx + c$ does not meet the $x$-axis |
|---|---|---|
| $ax^2 + bx + c = 0$ has **2 distinct real roots** | $ax^2 + bx + c = 0$ has **only 1 real root** (equal roots) | $ax^2 + bx + c = 0$ has **no real roots** |
| $b^2 - 4ac > 0$ **discriminant is positive** | $b^2 - 4ac = 0$ **discriminant is zero** | $b^2 - 4ac < 0$ **discriminant is negative** |

Calculating the discriminant allows you to determine the **nature** of the roots:

$b^2 - 4ac > 0$ —discriminant positive→ two distinct real roots

$b^2 - 4ac = 0$ —discriminant zero→ one real root (two equal roots)

$b^2 - 4ac < 0$ —discriminant negative→ no real roots

### Example

Determine the nature of the roots of these equations:

a) $x^2 - 4x + 3 = 0$ b) $x^2 - 4x + 4 = 0$ c) $x^2 - 4x + 5 = 0$

### Solution

In each case compare the equation with $ax^2 + bx + c = 0$

a) $a = 1$, $b = -4$, $c = 3$

$b^2 - 4ac = (-4)^2 - 4 \times 1 \times 3 = 4$

Discriminant is positive so there are **two distinct real roots**.

b) $a = 1$, $b = -4$, $c = 4$

$b^2 - 4ac = (-4)^2 - 4 \times 1 \times 4 = 0$

Discriminant is zero so there is **one real root** (equal roots).

c) $a = 1$, $b = -4$, $c = 5$

$b^2 - 4ac = (-4)^2 - 4 \times 1 \times 5 = -4$

Discriminant is negative so there are **no real roots**.

## Tangency

The discriminant can be used to show that lines are tangents to curves:

For example to find where a line $y = mx + d$ meets a parabola $y = ax^2 + bx + c$ you proceed as follows:

$$\left.\begin{aligned} y &= ax^2 + bx + c \\ y &= mx + d \end{aligned}\right\}$$

For points of intersection replace $y$ by $mx + d$ in the quadratic equation.

This will give a quadratic equation.

Calculate the discriminant of this equation. If it is **zero** then there is only **one solution**. This means **one point of intersection** and so the line is **a tangent** to the curve.

### Example

Show that $y = 4x - 2$ is a tangent to the parabola $y = x^2 + 2$

### Solution

To find the points of intersection solve:

$$\left.\begin{aligned} y &= x^2 + 2 \\ y &= 4x - 2 \end{aligned}\right\} \quad \text{so} \quad \begin{aligned} x^2 + 2 &= 4x - 2 \\ x^2 - 4x + 4 &= 0 \end{aligned}$$

Discriminant $= (-4)^2 - 4 \times 1 \times 4 = 0$

so there is one solution and hence one point of intersection. **The line $y = 4x - 2$ is a tangent.**

# Quadratic inequalities

If $f(x)$ is any quadratic expression like $ax^2 + bx + c$ then to solve quadratic inequalities like

$f(x) > $ or $f(x) < 0$

follow these steps:

**Step 1** Solve the corresponding quadratic equation $f(x) = 0$ to find the $x$-axis intercepts for the graph $y = f(x)$.

**Step 2** Sketch the graph $y = f(x)$ clearly showing the $x$-axis intercepts.

**Step 3** Write down the solution using the sketch of the graph $y = f(x)$. Examples are given in this table:

| Inequality: | What to look for: | 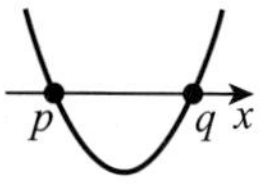 For this graph the solution is: (U-shaped graph crossing $x$-axis at $p$ and $q$) | 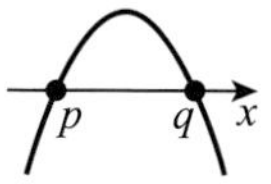 For this graph the solution is: (∩-shaped graph crossing $x$-axis at $p$ and $q$) |
|---|---|---|---|
| $ax^2 + bx + c > 0$ | Where is the graph **above** the $x$-axis? | $x < p$ or $x > q$ | $p < x < q$ |
| $ax^2 + bx + c < 0$ | Where is the graph **below** the $x$-axis? | $p < x < q$ | $x < p$ or $x > q$ |

Note: If the inequality uses the signs $\leq$ or $\geq$ then so will the solution.

## Example

Find the real values of $x$ satisfying $2x^2 + x - 1 > 0$

## Solution

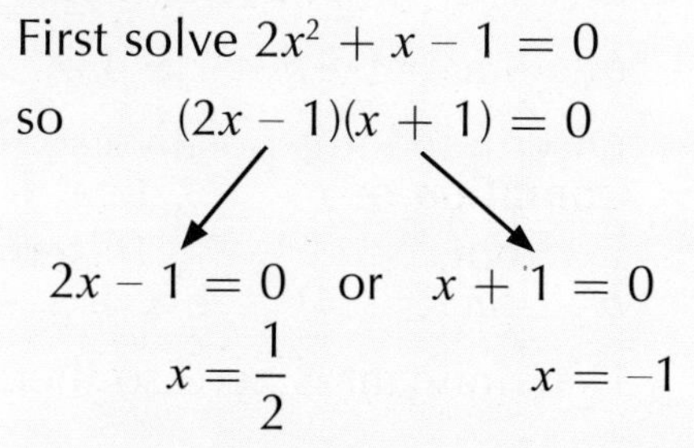

First solve $2x^2 + x - 1 = 0$

so $(2x - 1)(x + 1) = 0$

$2x - 1 = 0$ or $x + 1 = 0$

$x = \frac{1}{2}$ $\quad$ $x = -1$

**Sketch of $y = 2x^2 + x - 1$**

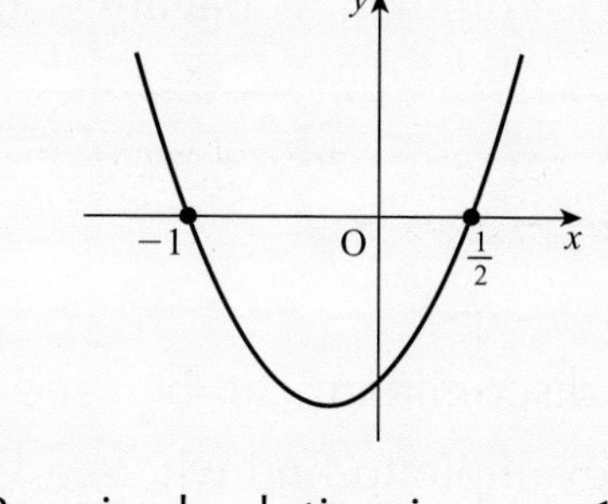


Required solution is $\boldsymbol{x < -1}$ **or** $\boldsymbol{x > \frac{1}{2}}$

(graph is **above** $x$-axis for these values).

# Quadratic theory revisited II

## Restrictions on the discriminant

It may be necessary to impose conditions on the nature of the roots of a quadratic equation. This can be done by applying restrictions to the discriminant. Here's how:

| Condition on roots | | Restriction on discriminant |
|---|---|---|
| Real ('one or two') | $\Longrightarrow$ | Positive or zero ($b^2 - 4ac \geq 0$) |
| Equal ('one') | $\Longrightarrow$ | Zero ($b^2 - 4ac = 0$) |
| Non-real ('none') | $\Longrightarrow$ | Negative ($b^2 - 4ac < 0$) |

**Example**

For what values of $p$ does $x^2 - 2x + p = 0$ have real roots?

**Solution**

Compare $x^2 - 2x + p = 0$

with $ax^2 + bx + c = 0$

This gives $a = 1$, $b = -2$ and $c = p$

Discriminant $= b^2 - 4ac$

$= (-2)^2 - 4 \times 1 \times p$

$= 4 - 4p$

The condition 'real roots' requires the discriminant to be restricted to positive values or zero.

So $4 - 4p \geq 0$ giving $-4p \geq -4$

so $\boldsymbol{p \leq 1}$

**Example**

Find values of $k$ so that $\frac{2(3x+1)}{3x^2+1} = k$ has two equal roots.

**Solution**

Rearrange the equation…

$6x + 2 = 3kx^2 + k$ giving $3kx^2 - 6x + k - 2 = 0$ and comparing this quadratic equation with $ax^2 + bx + c = 0$ gives $a = 3k$, $b = -6$, $c = k - 2$

Discriminant $= b^2 - 4ac = (-6)^2 - 4 \times 3k(k - 2)$

$= 36 - 12k^2 + 24k$

$= 36 + 24k - 12k^2$

The condition 'equal roots' gives the restriction 'discriminant = 0'.

So $36 + 24k - 12k^2 = 0$ giving $12(3 + 2k - k^2) = 0$ so $12(3 - k)(1 + k) = 0$

This gives two possible values for $k$ namely $\boldsymbol{k = 3}$ **or** $\boldsymbol{k = -1}$

## Building equations from the roots

It is possible to 'design' a quadratic equation that has two particular roots.
If the roots are $x = a$ and $x = b$ then $(x - a)(x - b) = 0$ is one such equation.

**Example**

Find a quadratic equation that has roots $\frac{1}{3}$ and $-2$

**Solution**

One such equation is

$$\left(x - \frac{1}{3}\right)(x + 2) = 0$$

This gives $x^2 + 2x - \frac{1}{3}x - \frac{2}{3} = 0$

Multiplying both sides of the equation by 3 gives:

$3x^2 + 6x - x - 2 = 0$

$$\mathbf{3x^2 + 5x - 2 = 0}$$

**TOP TIP**

This technique extends to cubic equations with three roots.

## Where are the roots?

The graph is below the $x$-axis at $x = a$ and above the $x$-axis at $x = b$. If the graph is a continuous curve then it must cross the $x$-axis somewhere between $a$ and $b$, at $x = \alpha$ say:

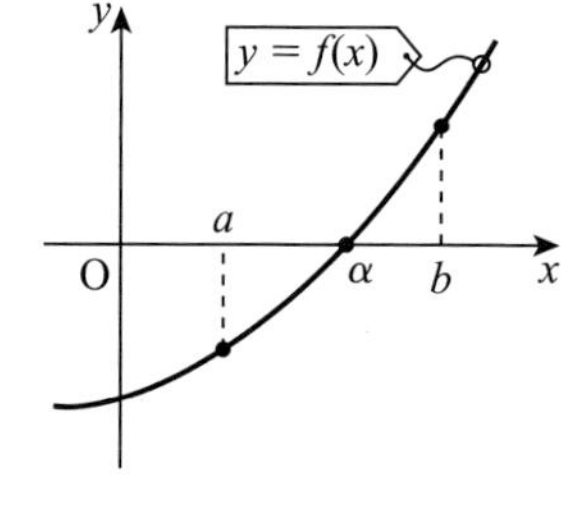


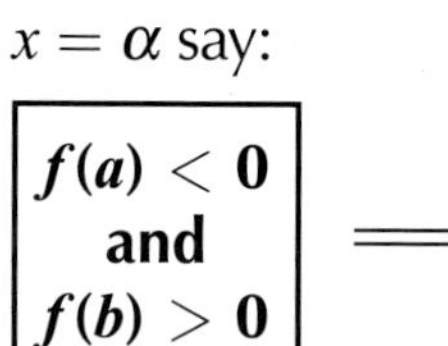


$\Longrightarrow$ **$f(\alpha) = 0$ for some value $\alpha$ with $a < \alpha < b$**

i.e. $f(x) = 0$ has a root between $a$ and $b$.

**Example**

Show that $x^3 - 3x + 1 = 0$ has a root between 1 and 2.

**Solution**

Let $f(x) = x^3 - 3x + 1$

then $f(1) = -1$ so $f(1) < 0$

and $f(2) = 3$ so $f(2) > 0$

Hence there is a root, $\alpha$ say, with $1 < \alpha < 2$.

**TOP TIP**

Zooming in on a graph using a graphing calculator is very useful for locating roots but you must always write down evidence in your written solution.

## Intersection of two parabolas

To find where two curves $y = f(x)$ and $y = g(x)$ intersect you solve the equation $f(x) = g(x)$

**Example**

Find the coordinates of the intersection points $A$ and $B$ as shown in the diagram.

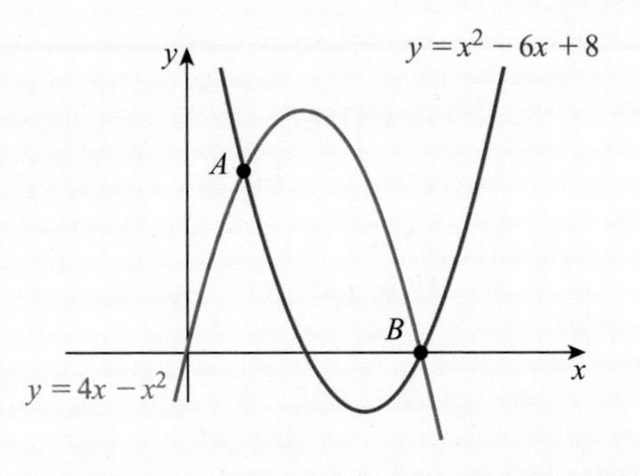


**Solution**

$x^2 - 6x + 8 = 4x - x^2$

$2x^2 - 10x + 8 = 0$

$x^2 - 5x + 4 = 0$

$(x - 1)(x - 4) = 0$

$x = 1$ or $x = 4$

Using $y = 4x - x^2$

when $x = 1$ $y = 4 \times 1 - 1^2 = 4 - 1 = 3$ so $A(1, 3)$

when $x = 4$ $y = 4 \times 4 - 4^2 = 16 - 16 = 0$ so $B(4, 0)$

# Quick Test 17

1. Use the discriminant to show that $y = 3x + 2$ is a tangent to the curve $y = x^2 - 11x + 51$
2. For what values of $k$ does $kx^2 - 3x + k = 0$ have real roots?
3. Solve $20 + x - x^2 > 0$
4. Find a quadratic equation with roots $x = -\frac{1}{3}$ and $x = \frac{1}{4}$
5. Two functions are defined by $f(x) = 3x^2 - x - 2$ and $g(x) = x^2 + 4x + 1$ with domain the set of real numbers. Find the intersection points of the graphs $y = f(x)$ and $y = g(x)$

# Polynomials and synthetic division

## What is a polynomial?

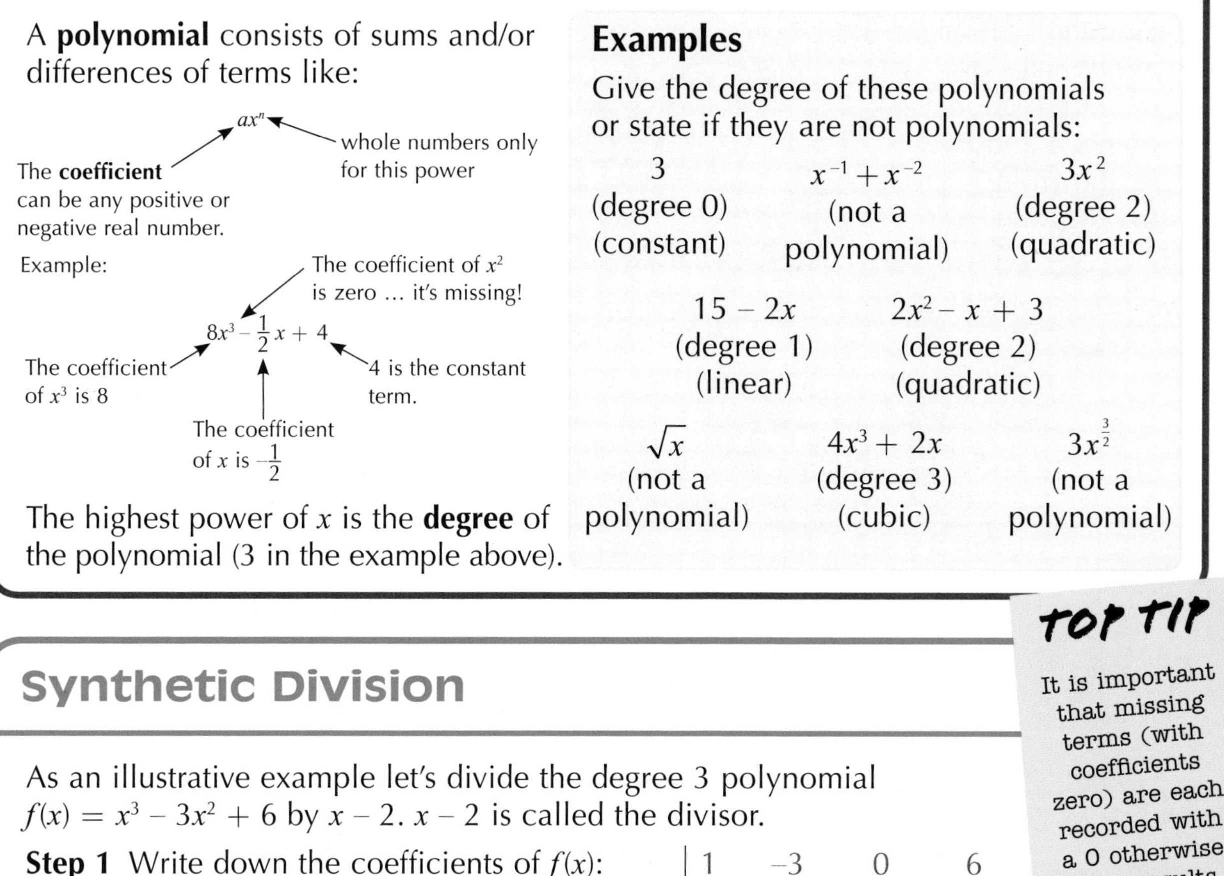

A **polynomial** consists of sums and/or differences of terms like:

$ax^n$

The **coefficient** can be any positive or negative real number.

whole numbers only for this power

Example:

$8x^3 - \frac{1}{2}x + 4$

The coefficient of $x^3$ is 8

The coefficient of $x^2$ is zero … it's missing!

The coefficient of $x$ is $-\frac{1}{2}$

4 is the constant term.

The highest power of $x$ is the **degree** of the polynomial (3 in the example above).

**Examples**

Give the degree of these polynomials or state if they are not polynomials:

| | | |
|---|---|---|
| $3$ (degree 0) (constant) | $x^{-1} + x^{-2}$ (not a polynomial) | $3x^2$ (degree 2) (quadratic) |
| $15 - 2x$ (degree 1) (linear) | $2x^2 - x + 3$ (degree 2) (quadratic) | |
| $\sqrt{x}$ (not a polynomial) | $4x^3 + 2x$ (degree 3) (cubic) | $3x^{\frac{3}{2}}$ (not a polynomial) |

**TOP TIP**

It is important that missing terms (with coefficients zero) are each recorded with a 0 otherwise your results will be wrong!

## Synthetic Division

As an illustrative example let's divide the degree 3 polynomial $f(x) = x^3 - 3x^2 + 6$ by $x - 2$. $x - 2$ is called the divisor.

**Step 1** Write down the coefficients of $f(x)$:
$f(x) = 1x^3 - 3x^2 + 0x + 6$

| 1 | –3 | 0 | 6 |

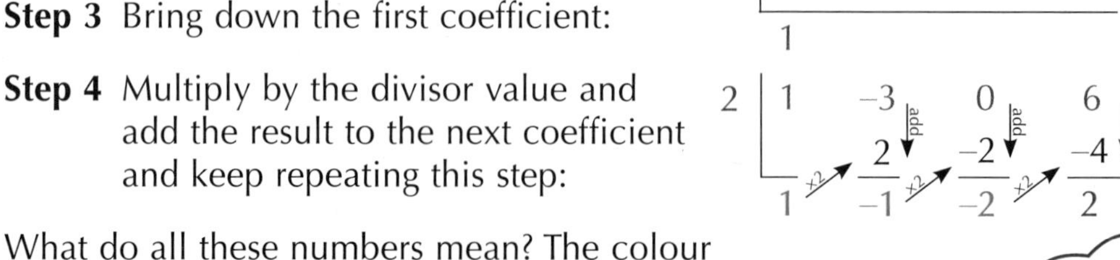

**Step 2** Find the value of $x$ that makes the divisor zero: 2

2 | 1 –3 0 6

**Step 3** Bring down the first coefficient:

2 | 1 –3 0 6
1

**Step 4** Multiply by the divisor value and add the result to the next coefficient and keep repeating this step:

2 | 1 –3 0 6
  2 –2 –4
1 –1 –2 2



What do all these numbers mean? The colour coding below explains how to interpret these numbers:

2 | 1 –3 0 6
  2 –2 –4
1 –1 –2 2

$f(x) = 1x^3 - 3x^2 + 0x + 6$
$= (x - 2)(1x^2 - 1x - 2) + 2$

Divisor | Quotient | Remainder

Compare this number example:
Divide 7 by 2: 2 | 7
3 rl
Giving: $7 = 2 \times 3 + 1$

## Using synthetic division

Here is a colour coded diagram showing $f(x)$ divided by $x - h$:

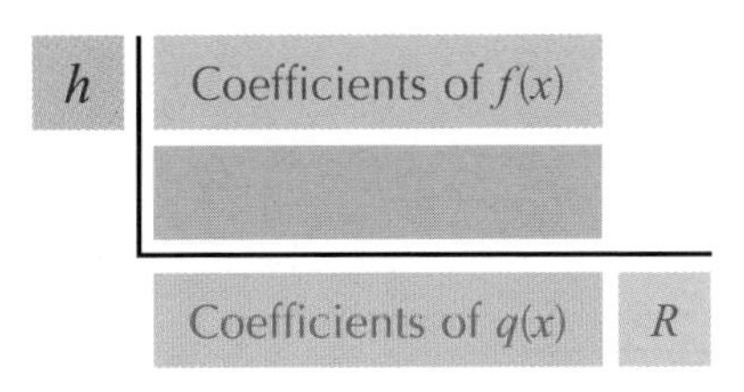


Giving: $f(x) = (x - h)q(x) + R$

There are different uses for synthetic division:

**Use 1** To calculate the value of $f(h)$. Notice that $f(h) = (h - h)q(x) + R = 0 + R = R$

**Use 2** To find the quotient $q(x)$ and remainder $R$ when $f(x)$ is divided by $x - h$

**Use 3** To find a factor $x - h$ of $f(x)$. This will happen if you find $R = 0$

**Example 1** $f(x) = 2x^3 - 3x + 10$ Find $f(-3)$

**Solution**

$$\begin{array}{r|rrrr} -3 & 2 & 0 & -3 & 10 \\ & & -6 & 18 & -45 \\ \hline & 2 & -6 & 15 & -35 \end{array}$$

So $f(-3) = -35$

**Example 2** $f(x) = 2x^3 - 3x + 10$
Find the quotient and remainder when $f(x)$ is divided by $x - 2$

**Solution**

$$\begin{array}{r|rrrr} 2 & 2 & 0 & -3 & 10 \\ & & 4 & 8 & 10 \\ \hline & 2 & 4 & 5 & 20 \end{array}$$

So $f(x) = (x - 2)(2x^2 + 4x + 5) + 20$

Quotient: $2x^2 + 4x + 5$
Remainder: 20

**Example 3** $f(x) = 2x^3 - 3x + 10$ Show that $x + 2$ is a factor of $f(x)$

**Solution**

$$\begin{array}{r|rrrr} -2 & 2 & 0 & -3 & 10 \\ & & -4 & 8 & -10 \\ \hline & 2 & -4 & 5 & 0 \end{array}$$

The remainder is 0 when $f(x)$ is divided by $x + 2$ so $x + 2$ is a factor of $f(x)$

**TOP TIP**
When dividing by $x + h$ you use $-h$ for the synthetic division

Notes:

- The fact that $f(x)$ is divided by $x - h$ $\Rightarrow R = f(h)$ is known as **The Remainder Theorem**
- The fact that $R = 0 \Leftrightarrow f(h)$ $x - h$ is a factor is known as **The Factor Theorem**

# Quick Test 18

1. Which of these are not polynomials? $x^{-1}$, $3x^2$, $\frac{1}{2}x$, $\frac{1}{2}\sqrt{x}$, $\frac{1}{2}$
2. Show that both $x - 3$ and $x + 2$ are factors of $x^4 - 7x^2 - 6x$
3. Find the quotient and remainder when $4x^3 - 12x + 7$ is divided by $x + 2$
4. Use synthetic division to calculate the exact value of $f(-\frac{1}{3})$ where $f(x) = x^4 + 2x^3 + x^2 + 1$ (no calculators!)

# Factors, roots and graphs

## Factorising polynomials

If you divide 6 by 3 the quotient is 2 and the remainder is 0. $6 = 3 \times 2$ so 3 is a factor of 6.

Similarly if you divide $f(x)$ by $x - h$ and find the remainder is 0 then $f(x) = (x - h) \times q(x)$ so $x - h$ is a factor of $f(x)$.

Using synthetic division:

| $h$ | Coefficients of $f(x)$ | |
|---|---|---|
| | | |
| | Coefficients of $q(x)$ | 0 |

$f(x) = (x - h) \times q(x)$

$(x - h)$ is a factor

You are using the Factor Theorem:

Remainder is zero. $\Longleftrightarrow$ $x - h$ is a factor.

Note: Be systematic when hunting for a value of $h$ that gives a zero remainder. Try 1 then −1 then 2 then −2 and so on.

**Example**

Factorise $f(x) = x^3 + 2x^2 - 5x - 6$

**Solution** Divide $f(x)$ by $x - 1$:

| 1 | 1 | 2 | −5 | −6 |
|---|---|---|---|---|
| | | 1 | 3 | −2 |
| | 1 | 3 | −2 | −8 |

remainder is not zero so $x - 1$ is not a factor

Now let's try $x + 1$ as a candidate:

| −1 | 1 | 2 | −5 | −6 |
|---|---|---|---|---|
| | | −1 | −1 | 6 |
| | 1 | 1 | −6 | 0 |

remainder is zero so $x + 1$ is a factor

So $f(x) = (x + 1)(x^2 + x - 6)$ ← quadratic factorising

$= (x + 1)(x - 2)(x + 3)$

**TOP TIP**

When hunting for factors look at the constant term of $f(x)$. You only need to try values of $h$ that are factors of this constant.

## Factors, roots and graphs

If $x - h$ is a factor of $f(x)$ then $h$ is a root of $f(x) = 0$ and the graph $y = f(x)$ cuts the $x$-axis at $(h, 0)$.

Example: Let $f(x) = x^3 + 2x^2 - 5x - 6$

$f(x) = (x + 3)(x + 1)(x - 2)$

The roots of $f(x) = 0$ are: −3 −1 2

**The graph $y = f(x)$**

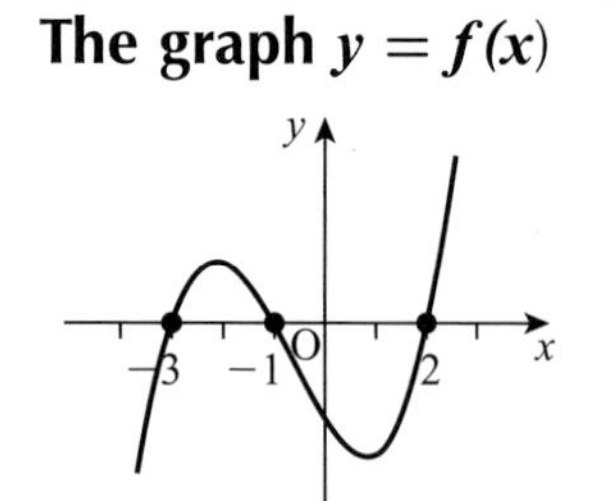


## Finding an unknown coefficient

**Example** $x^3 + 2x^2 + kx - 6$ has a factor of $x + 1$. Calculate the value of $k$.

**Solution**

| −1 | 1 | 2 | $k$ | −6 |
|---|---|---|---|---|
| | | −1 | −1 | $-k + 1$ |
| | 1 | 1 | $k - 1$ | $-k - 5$ |

Since $x + 1$ is a factor then the remainder will be 0.

So: $-k - 5 = 0 \Rightarrow -k = 5 \Rightarrow k = -5$

## Problem solving

**Example** $f(x) = 12x^3 - 28x^2 - 9x + 10$

a) Show that $\frac{1}{2}$ is a root of $f(x) = 0$ and find the other two roots.

b) Hence find where the graph $y = f(x)$ cuts the $x$-axis.

### a) Solution

Let $f(x) = 12x^3 - 28x^2 - 9x + 10$

Synthetic division gives:

| $\frac{1}{2}$ | 12 | −28 | −9 | 10 |
|---|---|---|---|---|
| | | 6 | −11 | −10 |
| | 12 | −22 | −20 | 0 |

Since $f\left(\frac{1}{2}\right) = 0$ then $\frac{1}{2}$ is a root.

Using the table above...

$$f(x) = \left(x - \tfrac{1}{2}\right)(12x^2 - 22x - 20)$$
$$= \left(x - \tfrac{1}{2}\right) \times 2(6x^2 - 11x - 10)$$
$$= (2x - 1)(6x^2 - 11x - 10)$$

so

$$f(x) = (2x - 1)(3x + 2)(2x - 5)$$

and so $f(x) = 0$ gives

$2x - 1 = 0$ or $3x + 2 = 0$ or $2x - 5 = 0$

$\boldsymbol{x = \frac{1}{2}}$ $\quad \boldsymbol{x = -\frac{2}{3}}$ $\quad \boldsymbol{x = \frac{5}{2}}$

### b) Solution

So $y = f(x)$ crosses the $x$-axis at the points:

$\left(\frac{1}{2}, 0\right)$, $\left(-\frac{2}{3}, 0\right)$ and $\left(\frac{5}{2}, 0\right)$

# Quick Test 19

1. a) Show that $x - 1$ is a factor of $f(x) = 2x^4 - 3x^3 - x^2 + 3x - 1$.
   b) Hence factorise $f(x)$ into two factors.
   c) Now show that $x + 1$ is a factor of one of these factors.
   d) Hence express $f(x)$ in fully factorised form.
2. Find where the graph $y = f(x)$ crosses the $x$-axis if $f(x) = 6x^3 - 7x^2 - x + 2$

**TOP TIP**

To factorise **fully** you must break up any quadratic or cubic factors into further factor pairs if possible.

It's like $30 = 2 \times 15$ where 15 can be broken into $3 \times 5$ so $30 = 2 \times 3 \times 5$ and is now **fully factorised.**

# Solving trig equations

## Solving trig equations using radians

**Step 1** Rearrange the equation to one of the forms:
$\sin\theta = k$ or $\cos\theta = k$ or $\tan\theta = k$
where $k$ is a constant (positive or negative)

**Step 2** Which quadrants can $\theta$ be in ? Use the sign of $k$ (positive or negative) and this diagram:

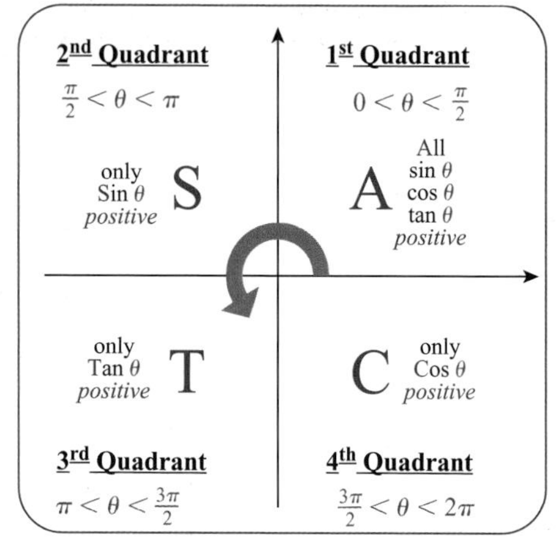


**Step 3** Find the related 1st quadrant angle $\alpha$. To find this angle you assume $k$ is positive and solve the equation for the 1st quadrant.

**Step 4** Now use the value of $\alpha$ to find the possible values for $\theta$ knowing the quadrants from Step 2 above:

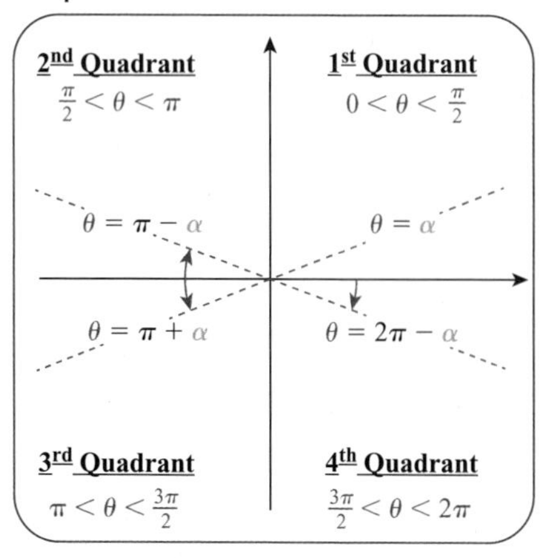


### Example 1

Solve: $2\sin x + 1 = 0,\ 0 \le x < 2\pi$

### Solution

1. $2\sin x + 1 = 0$ rearranges to $\sin x = -\frac{1}{2}$
2. $\sin x$ is negative for angles in the 3rd or 4th quadrants
3. 1st quadrant angle is $\frac{\pi}{6}$ (since $\sin\frac{\pi}{6} = \frac{1}{2}$)
4. $x = \pi + \frac{\pi}{6}$ (3rd quadrant) or $x = 2\pi - \frac{\pi}{6}$ (4th quadrant)

   $x = \frac{6\pi}{6} + \frac{\pi}{6}$ or $x = \frac{12\pi}{6} - \frac{\pi}{6}$

   $x = \frac{7\pi}{6}$ or $x = \frac{11\pi}{6}$

Note:
A graphical check may be made using the 'sine graph':

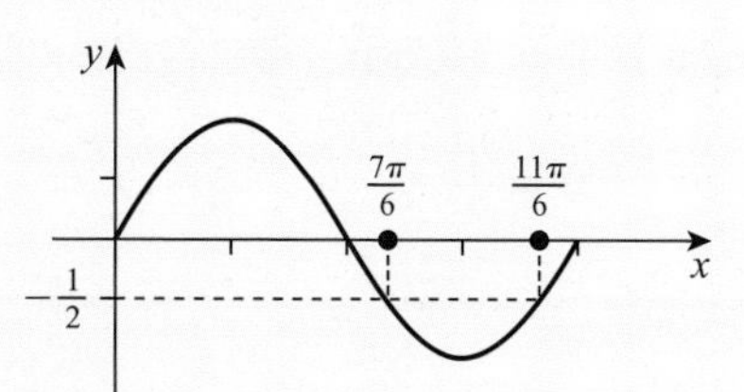


The diagram indicates the two solutions to $\sin x = -\frac{1}{2}$.

### Example 2

Solve $3\cos(2x - \frac{\pi}{6}) = 2$ for $0 \le x < \pi$

**TOP TIP**
Pay close attention to the allowed values for $x$.

### Solution

Rearrange to $\cos(2x - \frac{\pi}{6}) = \frac{2}{3}$

So the angle $2x - \frac{\pi}{6}$ is in the 1st or 4th quadrants since cosine is positive in these quadrants.

Notes:

- If $k = 0$, $-1$ or $1$ use the trig graphs to solve the equations $\sin\theta = k$ or $\cos\theta = k$
- Remember the **exact values** (see page 24)
- Make sure your calculator is in radian mode for radian work.

Set calculator to Radian Mode. Enter $\frac{2}{3}$ and $\boxed{\cos^{-1}}$ giving 0·841 radians as the 1st quadrant angle.

so $2x - \frac{\pi}{6} = 0{\cdot}841$ (1st quad) or $2x - \frac{\pi}{6} = 2\pi - 0{\cdot}841$ (4th quad)

giving $2x = 0{\cdot}841 + \frac{\pi}{6} = 1{\cdot}364...$ or $2x = 2\pi - 0{\cdot}841 + \frac{\pi}{6} = 5{\cdot}965...$

so $\mathbf{x = 0{\cdot}682}$ or $\mathbf{x = 2{\cdot}983}$ (to 3 dec. places)

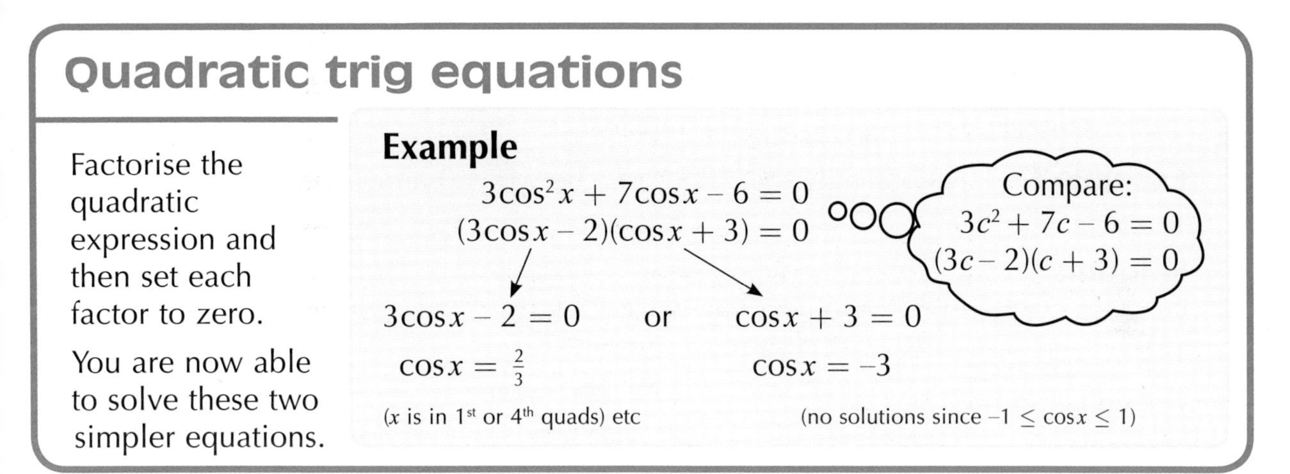

## Quadratic trig equations

Factorise the quadratic expression and then set each factor to zero.

You are now able to solve these two simpler equations.

**Example**

$$3\cos^2 x + 7\cos x - 6 = 0$$
$$(3\cos x - 2)(\cos x + 3) = 0$$

Compare:
$3c^2 + 7c - 6 = 0$
$(3c - 2)(c + 3) = 0$

$3\cos x - 2 = 0$ or $\cos x + 3 = 0$

$\cos x = \frac{2}{3}$ ($x$ is in 1st or 4th quads) etc

$\cos x = -3$ (no solutions since $-1 \le \cos x \le 1$)

## Double angle formulae and equations

Here is the general method to solve a trig equation in which a double angle appears:

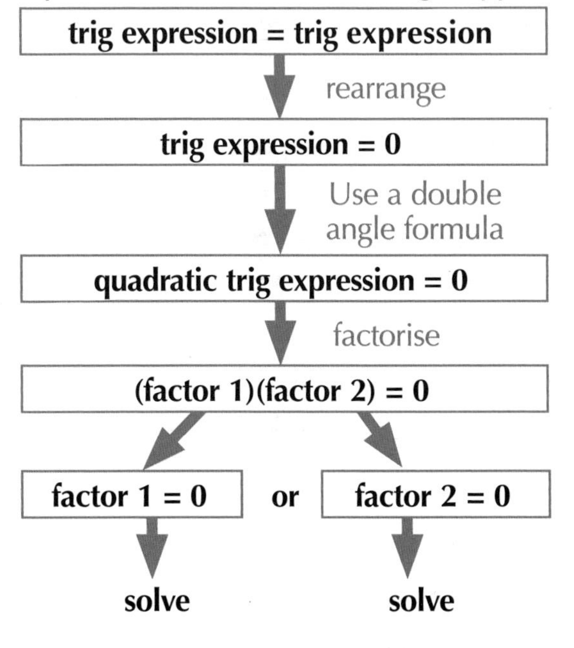


**Example 1**

Solve $\cos 2x^\circ + 3 = 5(1 + \cos x^\circ)$ for $0 \le x \le 360$

**Solution**

$$\cos 2x^\circ + 3 = 5 + 5\cos x^\circ$$
$$\cos 2x^\circ - 5\cos x^\circ - 2 = 0$$
$$2\cos^2 x^\circ - 1 - 5\cos x^\circ - 2 = 0$$
$$2\cos^2 x^\circ - 5\cos x^\circ - 3 = 0$$

(a quadratic in $\cos x^\circ$)

$$(2\cos x^\circ + 1)(\cos x^\circ - 3) = 0$$

$2\cos x^\circ + 1 = 0$ or $\cos x^\circ - 3 = 0$

$2\cos x^\circ = -1$, $\cos x^\circ = -\frac{1}{2}$ ($x^\circ$ is in 2nd or 3rd quads) (1st quad angle is 60°)

$\cos x^\circ = 3$, no solutions (since $-1 \le \cos x^\circ \le 1$)

so $x^\circ = 180^\circ - 60^\circ$ or $180^\circ + 60^\circ$

$\mathbf{x^\circ = 120^\circ}$ **or** $\mathbf{240^\circ}$

# Equations involving the wave function

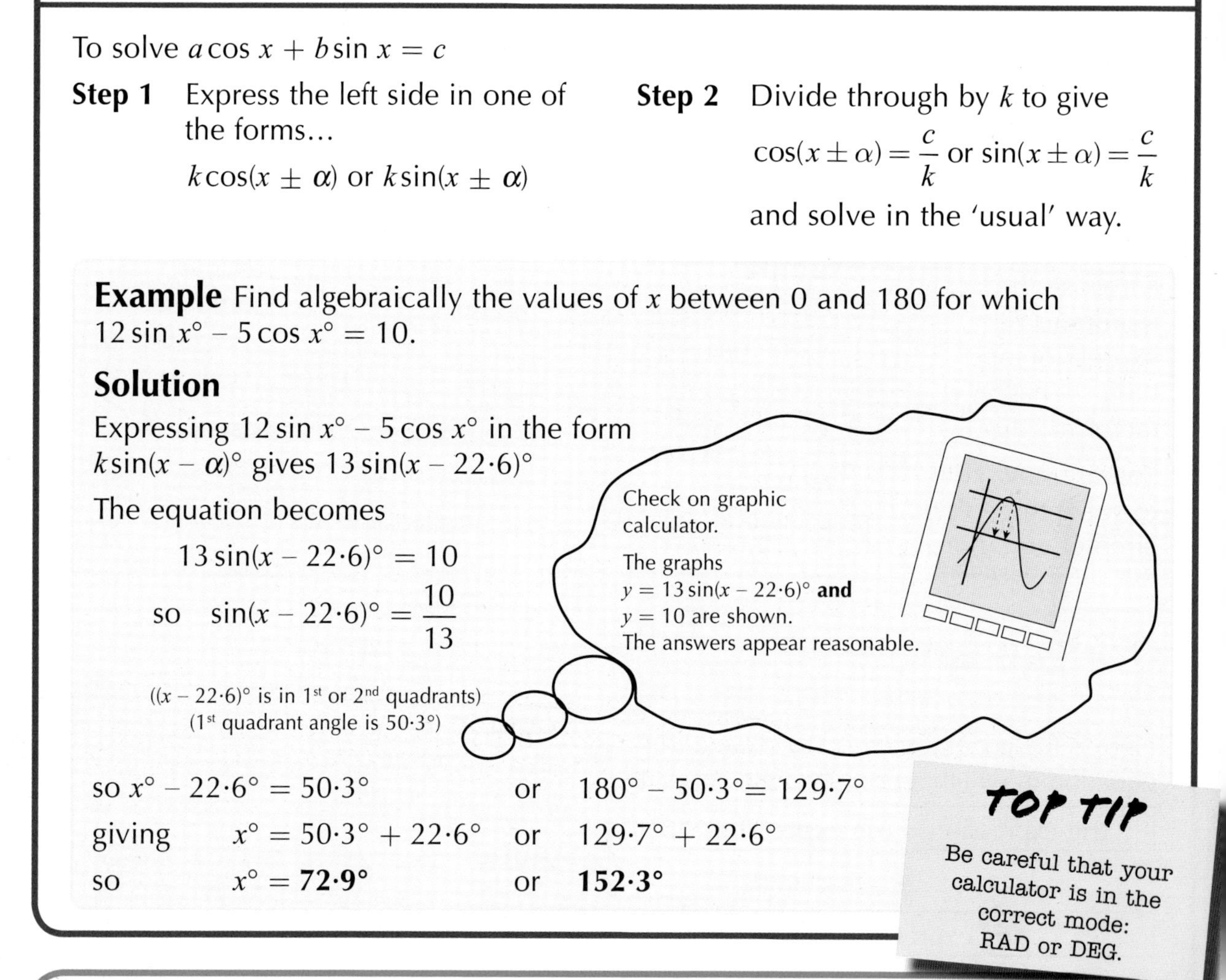

## Wave functions equations

To solve $a\cos x + b\sin x = c$

**Step 1** Express the left side in one of the forms...

$k\cos(x \pm \alpha)$ or $k\sin(x \pm \alpha)$

**Step 2** Divide through by $k$ to give

$\cos(x \pm \alpha) = \dfrac{c}{k}$ or $\sin(x \pm \alpha) = \dfrac{c}{k}$

and solve in the 'usual' way.

**Example** Find algebraically the values of $x$ between 0 and 180 for which $12\sin x^\circ - 5\cos x^\circ = 10$.

**Solution**

Expressing $12\sin x^\circ - 5\cos x^\circ$ in the form $k\sin(x - \alpha)^\circ$ gives $13\sin(x - 22{\cdot}6)^\circ$

The equation becomes

$$13\sin(x - 22{\cdot}6)^\circ = 10$$

$$\text{so}\quad \sin(x - 22{\cdot}6)^\circ = \frac{10}{13}$$

($(x - 22{\cdot}6)^\circ$ is in 1st or 2nd quadrants)
(1st quadrant angle is $50{\cdot}3^\circ$)

Check on graphic calculator.

The graphs $y = 13\sin(x - 22{\cdot}6)^\circ$ **and** $y = 10$ are shown.
The answers appear reasonable.

so $x^\circ - 22{\cdot}6^\circ = 50{\cdot}3^\circ$ or $180^\circ - 50{\cdot}3^\circ = 129{\cdot}7^\circ$

giving $x^\circ = 50{\cdot}3^\circ + 22{\cdot}6^\circ$ or $129{\cdot}7^\circ + 22{\cdot}6^\circ$

so $x^\circ = \mathbf{72{\cdot}9^\circ}$ or $\mathbf{152{\cdot}3^\circ}$

**TOP TIP**

Be careful that your calculator is in the correct mode: RAD or DEG.

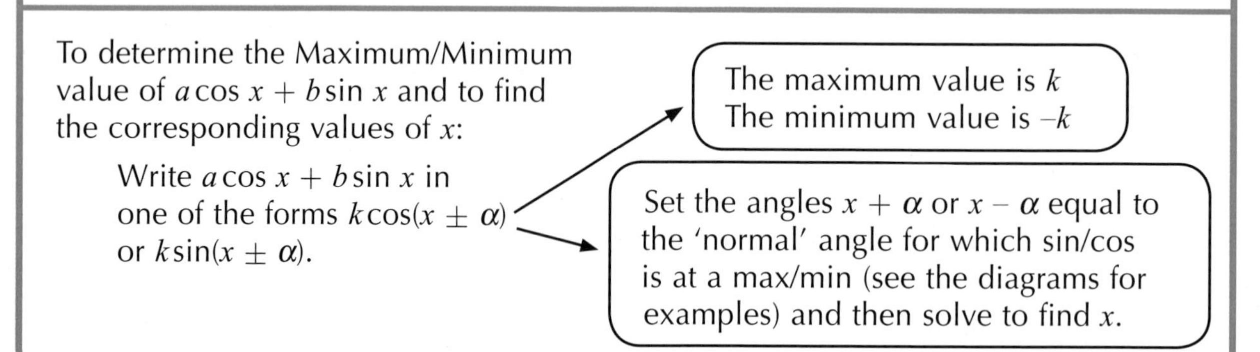

## Maximum and minimum values

To determine the Maximum/Minimum value of $a\cos x + b\sin x$ and to find the corresponding values of $x$:

Write $a\cos x + b\sin x$ in one of the forms $k\cos(x \pm \alpha)$ or $k\sin(x \pm \alpha)$.

The maximum value is $k$
The minimum value is $-k$

Set the angles $x + \alpha$ or $x - \alpha$ equal to the 'normal' angle for which sin/cos is at a max/min (see the diagrams for examples) and then solve to find $x$.

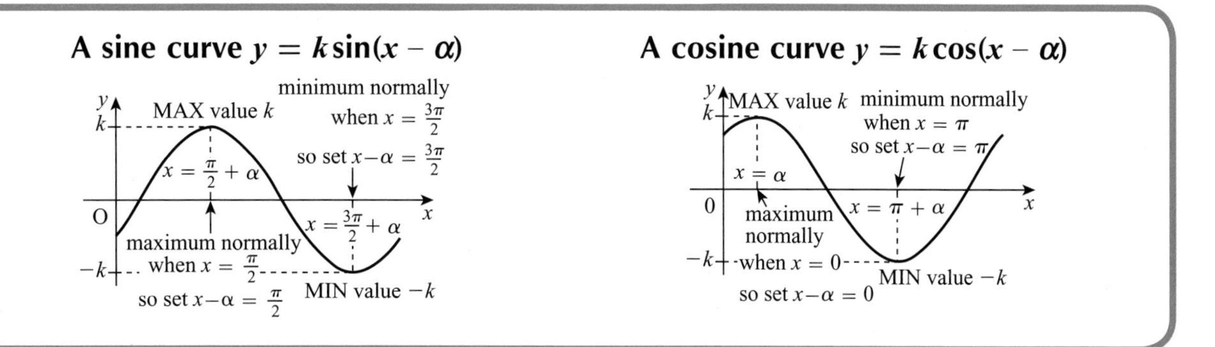


## Problem solving with the wave function

### Example 1

Find the maximum value of L where $L = \cos 2\theta + \sqrt{3}\sin 2\theta$ and the corresponding values for $\theta$ where $0 \leq \theta \leq 2\pi$.

### Solution

$\cos 2\theta + \sqrt{3}\sin 2\theta$ can be expressed as $k\cos(2\theta - \alpha)$ giving $k = 2$ and $\alpha = \frac{\pi}{3}$ so $L = 2\cos\left(2\theta - \frac{\pi}{3}\right)$

The maximum value of L is **2** and this happens when $2\theta - \frac{\pi}{3} = 0, 2\pi, \cdots$

$$\Rightarrow \quad 2\theta = \frac{\pi}{3}, 2\pi + \frac{\pi}{3}, \cdots$$

$$\Rightarrow \quad 2\theta = \frac{\pi}{3}, \frac{7\pi}{3}, \cdots$$

$$\Rightarrow \quad \theta = \frac{\pi}{6}, \frac{7\pi}{6}, \cdots$$

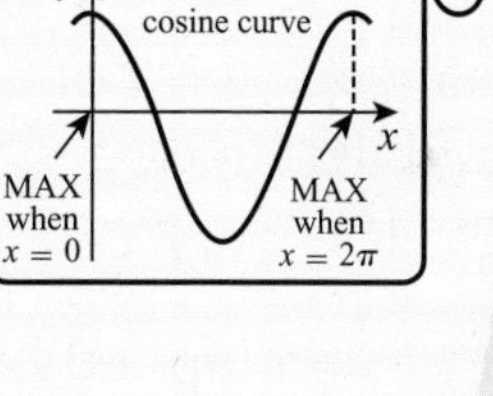


So $\theta = \frac{\pi}{6}$ **and** $\theta = \frac{7\pi}{6}$ are the required values (all other values are outside the range $0 \leq \theta \leq 2\pi$).

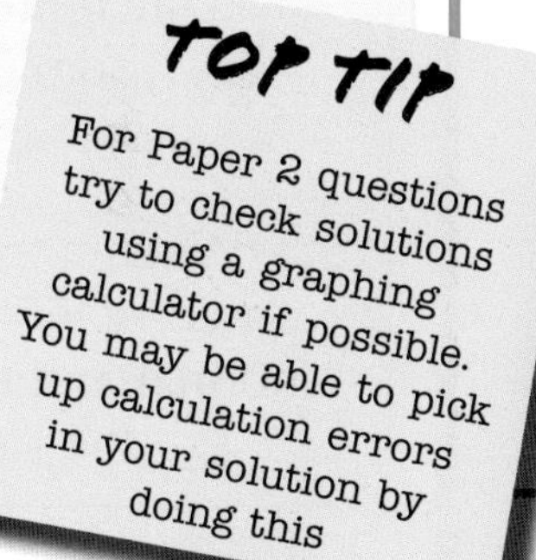


## Quick Test 20

1. Solve these equations
   a) $3\tan x° + 4 = 0$ for $0 \leq x \leq 360$
   b) $3\sin x + 1 = 0$ for $0 \leq x \leq 2\pi$
2. Find the exact solutions of these equations for $0 \leq x \leq 2\pi$:
   a) $\sqrt{3}\tan x = 1$  b) $2\cos x + 1 = 0$
3. Find algebraically the coordinates of $P$ and $Q$, the points of intersection of the two graphs.

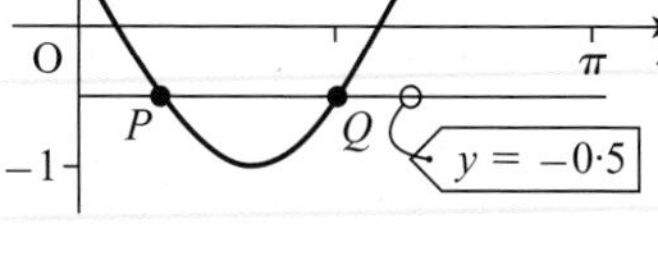


4. Solve: a) $4\cos^2 x° + 4\cos x° - 3 = 0$ for $0 \leq x \leq 360$ (to 1 decimal place)
   b) $3\sin x + 2 = \cos 2x$ for $0 \leq x \leq 2\pi$
5. $g(x) = \sqrt{3}\sin x° - \cos x°$
   a) Express $g(x)$ in the form $k\sin(x - \alpha)°$ where $k > 0$ and $0 \leq \alpha < 360$.
   b) Hence solve algebraically $g(x) = 0{\cdot}8$ for $0 \leq x < 360$.

# Basic rules and techniques

## What is differentiation?

The gradient at a point on a graph is given by the gradient of the tangent line at that point:

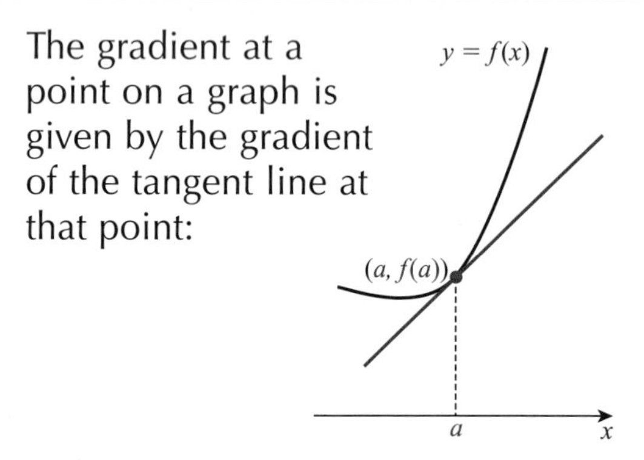


By calculating the gradient at every point on the graph $y = f(x)$ you have **differentiated** the function $f$ and produced a new gradient function $f'$ derived from $f$:

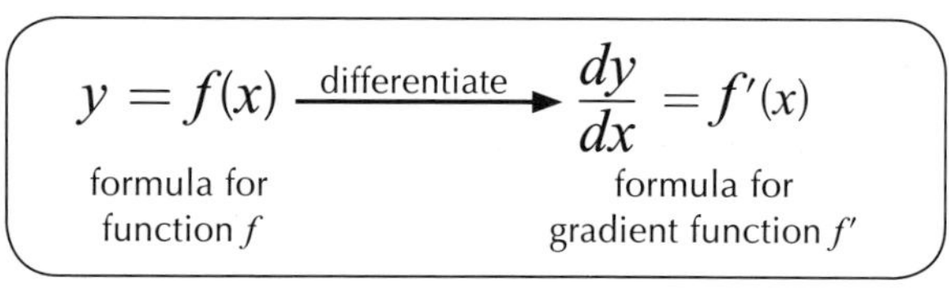


## The basic rules

**TOP TIP**

You should thoroughly revise the laws of indices as they are essential when differentiating expressions with powers.

| $f(x)$ | $f'(x)$ |
|---|---|
| $x^n$ | $nx^{n-1}$ |
| $g(x) \pm h(x)$ | $g'(x) \pm h'(x)$ |
| (Differentiate each term of a sum or difference.) | |
| $ag(x)$ | $ag'(x)$ |
| (When a term is multiplied by a constant then differentiate as normal and multiply the result by the same constant.) | |

**Special cases:**

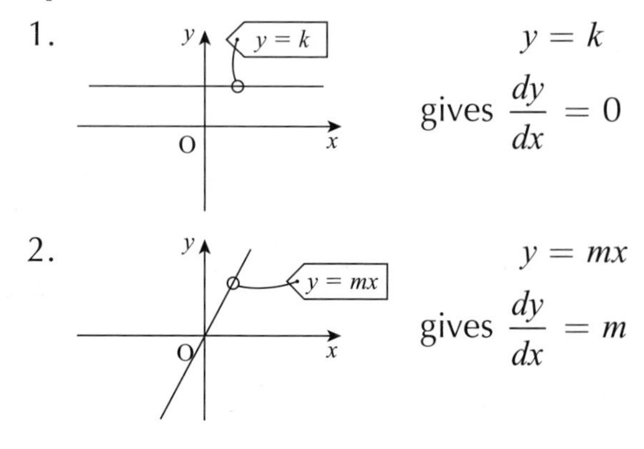


1. $y = k$ gives $\frac{dy}{dx} = 0$

2. $y = mx$ gives $\frac{dy}{dx} = m$

### Example

Differentiate:

a) $y = 5x^3 - 3x^2$

**Solution**

Differentiate each term.

$$\frac{dy}{dx} = 5 \times 3x^2 - 3 \times 2x^1$$

$$= \mathbf{15x^2 - 6x}$$

b) $f(x) = \frac{2}{\sqrt{x}} - \frac{3}{x}$

**Solution**

First prepare the 'formula' for differentiating by writing it as a difference of powers of $x$:

$$f(x) = \frac{2}{x^{\frac{1}{2}}} - \frac{3}{x^1} = 2x^{-\frac{1}{2}} - 3x^{-1}$$

Now use the differentiation rules:

$$f'(x) = 2 \times \left(-\frac{1}{2}\right)x^{-\frac{1}{2}-1} - 3 \times (-1)x^{-1-1}$$

$$= -x^{-\frac{3}{2}} + 3x^{-2}$$

# Differentiation techniques

You need to first 'prepare' an expression before attempting to differentiate it. Your aim is to write the expression as sums and/or differences of terms of the form $ax^n$

**technique 1** Remove root signs

e.g. $\sqrt{x} = x^{\frac{1}{2}}$ $\quad \dfrac{1}{\sqrt{x}} = x^{-\frac{1}{2}}$

**technique 2** Remove brackets

e.g. $(2x - 1)(x + 2)$
$= 2x^2 + 3x - 2$

**technique 3** Fractions with a single term on the denominator can be split:

e.g. $\dfrac{x^3 + x - 1}{x^2} = \dfrac{x^3}{x^2} + \dfrac{x}{x^2} - \dfrac{1}{x^2}$

$= x + x^{-1} - x^{-2}$
(using the Laws of Indices)

### Example 1

Calculate the **exact** value of $f'(9)$ where

$$f(x) = \frac{x - 3x^2}{\sqrt{x}}$$

### Solution

$$f(x) = \frac{x}{x^{\frac{1}{2}}} - \frac{3x^2}{x^{\frac{1}{2}}} = x^{\frac{1}{2}} - 3x^{\frac{3}{2}}$$

So $f'(x) = \dfrac{1}{2}x^{-\frac{1}{2}} - \dfrac{9}{2}x^{\frac{1}{2}} = \dfrac{1}{2\sqrt{x}} - \dfrac{9\sqrt{x}}{2}$

giving $f'(9) = \dfrac{1}{2\sqrt{9}} - \dfrac{9\sqrt{9}}{2} = \dfrac{1}{6} - \dfrac{27}{2}$

$= \dfrac{1}{6} - \dfrac{81}{6} = -\dfrac{80}{6} = -\mathbf{\dfrac{40}{3}}$

A decimal approximation is not acceptable for the **exact** value.

### Example 2

Find the gradient of the curve $y = x^3$ at the point $P(1, 1)$.

### Solution

$\dfrac{dy}{dx} = 3x^2$

When $x = 1$

$\dfrac{dy}{dx} = 3 \times 1^2 = 3$

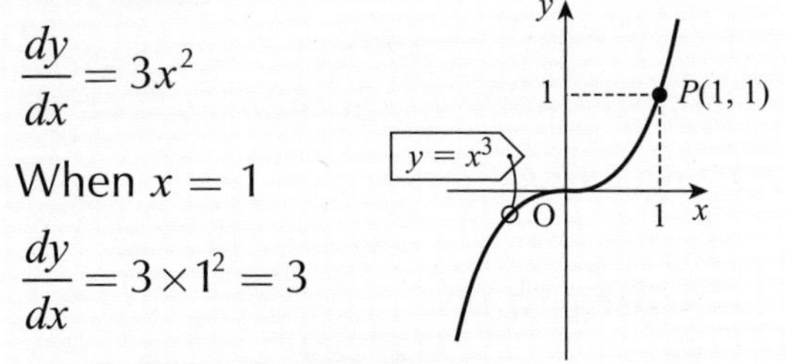


The required **gradient is 3**.

### Example 3

Find the points on the curve $y = x^3$ where the gradient is 12.

### Solution

You require $\dfrac{dy}{dx} = 12$ so $3x^2 = 12$, $x^2 = 4$

giving $x = 2$ or $-2$

Now substitute into $y = x^3$ for $y$-coordinates.

The required points are **(2, 8) and (–2, –8).**

The laws of indices tell you that $x^{\frac{3}{2}}$ means: find the square root of $x$ then cube the result

# Quick Test 21

1. Find $\frac{dy}{dx}$ where: a) $y = -\dfrac{1}{x^3}$ b) $y = \dfrac{2}{\sqrt{x}}$ c) $y = -\dfrac{1}{\sqrt{x}}$
2. Differentiate: a) $y = \dfrac{3x}{\sqrt{x}} - \dfrac{2}{x}$ b) $f(x) = \dfrac{2x + 1}{\sqrt{x}}$
3. Find: a) $f'(4)$ where $f(x) = 5\sqrt{x} - x$ b) $f'(9)$ where $f(x) = \dfrac{x^2 - x}{\sqrt{x}}$
4. Find the points on the graph of $y = \frac{1}{2}x^4$ where the gradient is –2

# Tangents and stationary points

## Equations of tangents

The gradient of the tangent to the graph $y = f(x)$ at the point $(x, y)$ is given by

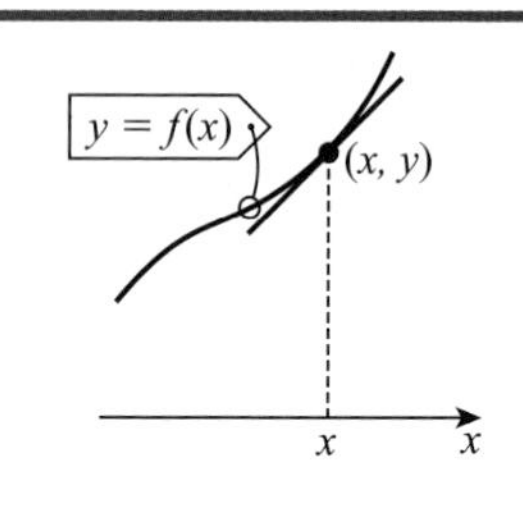


$$\frac{dy}{dx} \text{ or } f'(x)$$

In particular, at the point $(a, b)$ the gradient of the tangent is $f'(a)$.

To find the equation of a tangent line at the point $(a, b)$ on the graph $y = f(x)$:

**Step 1** Find $f'(x)$

**Step 2** Calculate $m = f'(a)$

**Step 3** Equation is $y - b = m(x - a)$

### Example

Find the equation of the tangent to $y = x^2$ at the point (3, 9).

### Solution

$\frac{dy}{dx} = 2x$. When $x = 3$ $\frac{dy}{dx} = 2 \times 3 = 6$

Gradient = 6. Point is (3,9).

Equation is $y - 9 = 6(x - 3)$

=> $y = 6x - 9$

**TOP TIP**

If you are asked for a **stationary point** then give the coordinates. If you are asked for a **stationary value** give the $y$-coordinate only.

## Finding stationary points

Points on a graph $y = f(x)$ where the gradient is zero are called **stationary points**.

$$f'(a) = 0$$

| $(a, f(a))$ is a stationary **point**. | $f(a)$ is a stationary **value**. |
|---|---|

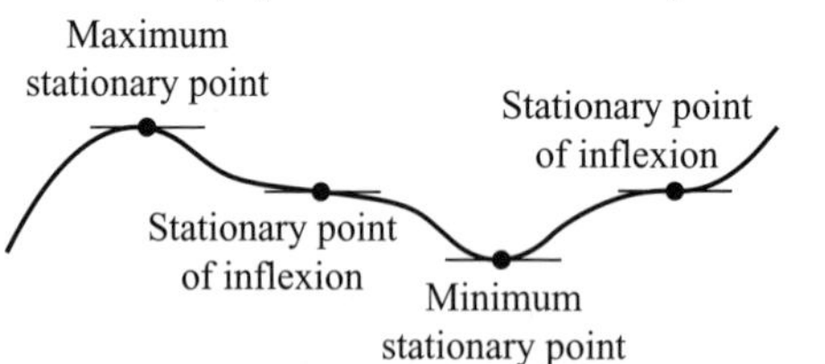


To find the stationary points on the graph $y = f(x)$:

**Step 1** Find $f'(x)$

**Step 2** Set $f'(x) = 0$

**Step 3** Solve $f'(x) = 0$

Each solution $x = a$ gives a stationary point.

**Step 4** Calculate $y = f(a)$ for each solution $x = a$. $(a, f(a))$ is a stationary point.

### Example

Find the stationary points on the graph

$$y = x^4 - 4x^3 + 3$$

### Solution

$y = x^4 - 4x^3 + 3$ gives $\frac{dy}{dx} = 4x^3 - 12x^2$

To find stationary points, set $\frac{dy}{dx} = 0$

so $4x^3 - 12x^2 = 0$

Now solve: $4x^2(x - 3) = 0$

$x^2 = 0$ or $x - 3 = 0$

$x = 0$ $\quad x = 3$

when $x = 0$ $\quad y = 0^4 - 4 \times 0^3 + 3 = 3$ giving (0, 3)

when $x = 3$ $\quad y = 3^4 - 4 \times 3^3 + 3 = -24$ giving (3, –24)

So there are two stationary points, namely (0, 3) and (3, –24), on this graph.

## Identifying types of stationary points

There are three types of stationary points:

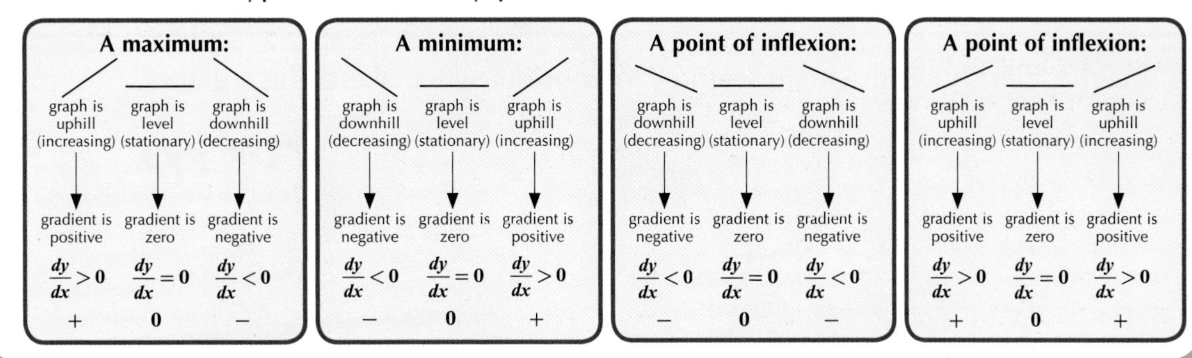


## Table of signs

Having identified where the stationary points are, say, at $x = a$ and $x = b$, then:

**Step 1** Draw a number line and place the $x$-values, $x = a$ and $x = b$, in order on the number line.

**Step 2** Underneath draw a two-row table:

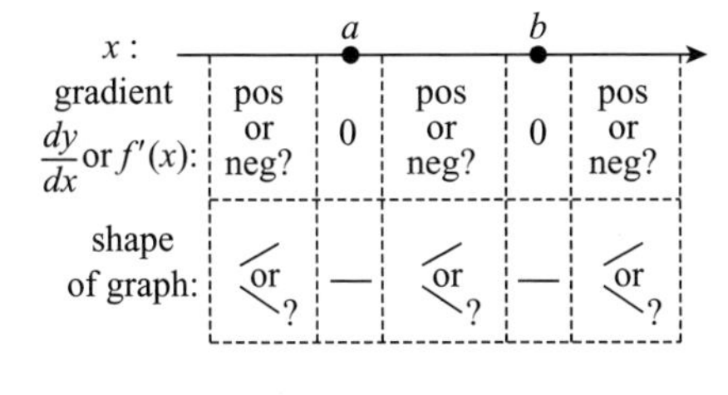


### Example

Determine the **nature** of the stationary points from the example at the bottom of page 60.

### Solution

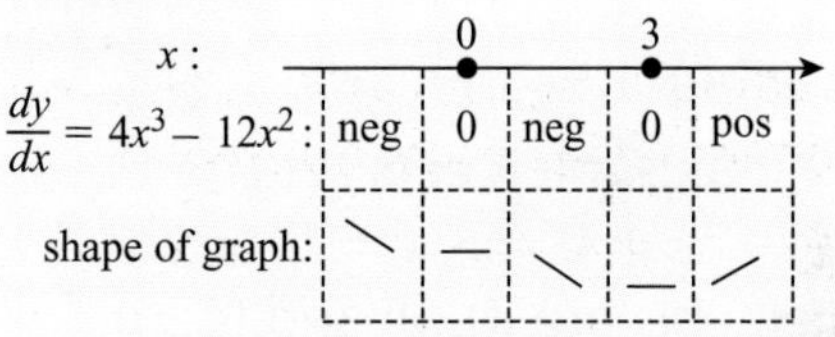


So (0, 3) is a stationary point of inflexion and (3, –24) is a minimum stationary point.

# Quick Test 22

When asked for the **nature** of stationary points you must give evidence. This should be a 'table of signs'.

1. On the curve $y = \frac{1}{3}x^3 - 2x + 1$ tangets are drawn at the points $A(-1, \frac{8}{3})$, $B(0, 1)$ and $C(1, -\frac{2}{3})$.

   a) Which pair of tangents are parallel?
   **Hint: Find their gradients.**

   b) Find the equations of the three tangents.

2. Find algebraically the coordinates of the stationary points and determine their nature for the curve: $y = 4x^5 + 5x^4 - 2$

# Graph sketching and gradients

## Sketching a graph

This diagram shows the main features to consider when sketching a graph:

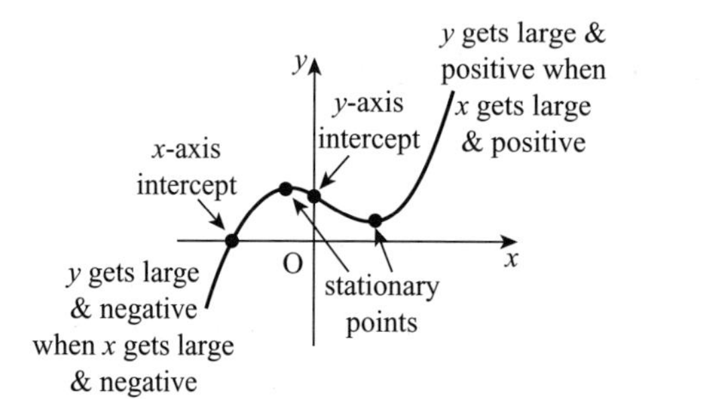


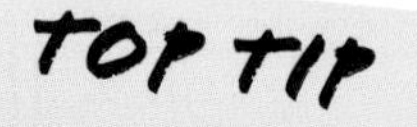


Remember you will only gain marks in the exam for this type of graphical work if you show evidence of your calculations.

When you want to sketch the graph $y = f(x)$ follow these steps:

**Step 1** Find where the graph cuts the $x$-axis: Set $y = 0$ and solve $f(x) = 0$.

**Step 2** Find where the graph cuts the $y$-axis: Set $x = 0$ and find the value of $f(0)$.

**Step 3** Find the stationary points and determine their nature:

Solve the equation $\frac{dy}{dx} = 0$ and then set up a table of signs.

**Step 4** Check the behaviour of $y$ for large positive/negative $x$ values.

### Example

Sketch the graph $y = x^3(4 - x)$

**Step 1** For $x$-axis intercepts set $y = 0$:

$x^3(4 - x) = 0 \Longrightarrow x = 0$ or $x = 4$

Intercepts are (0, 0) and (4, 0)

**Step 2** For $y$-axis intercepts set $x = 0$:

$y = 0^3 \times (4 - 0) = 0$

Intercept is (0, 0)

**Step 3** $y = x^3(4 - x) = 4x^3 - x^4$

$$\Longrightarrow \frac{dy}{dx} = 12x^2 - 4x^3 = 4x^2(3 - x)$$

For stationary points set $\frac{dy}{dx} = 0$

So $4x^2(3 - x) = 0 \Longrightarrow x = 0$ or $x = 3$

When $x = 3$ $y = 3^3 \times (4 - 3) = 27$

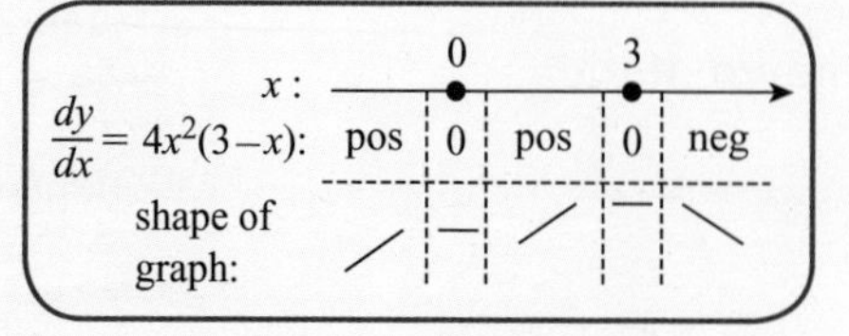

| $x$: | | 0 | | 3 | |
|---|---|---|---|---|---|
| $\frac{dy}{dx} = 4x^2(3-x)$: | pos | 0 | pos | 0 | neg |
| shape of graph: | / | — | / | — | \ |

(0, 0) is a point of inflexion

(3, 27) is maximum stationary point

Here is a diagram showing the information from steps 1, 2 and 3:

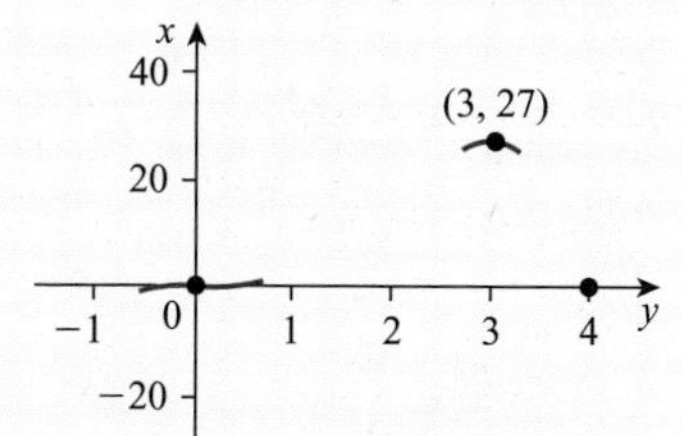


Complete the graph by joining up the known points with a smooth curve:

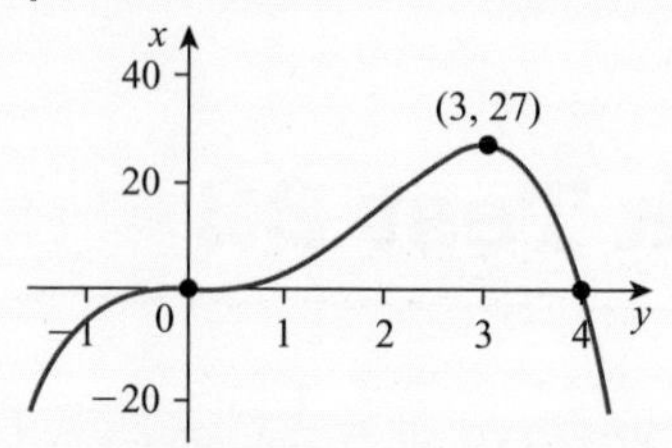


**Step 4** Check that as $x$ gets large and positive $y = x^3(4 - x)$ gets large and negative (try $x = 100$). Check that as $x$ gets large and negative $y = x^3(4 - x)$ gets large and negative (try $x = -100$).

# Sketching a gradient graph

You should be able to sketch the graph $\frac{dy}{dx} = f'(x)$ if you are given the graph $y = f(x)$

Here is a typical cubic function graph lined up with its gradient graph:

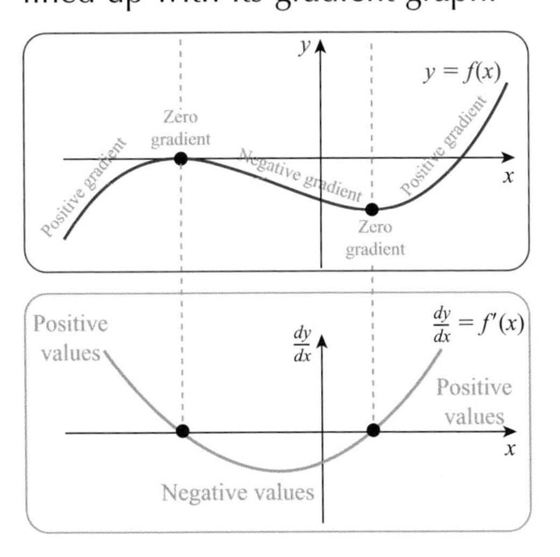


Notes:

- The graph $y = f(x)$ shows the values of the function $f$.
  The graph $\frac{dy}{dx} = f'(x)$ shows the values of the gradient of the original graph.
- Differentiating a cubic (degree 3) will give a quadratic (degree 2) which explains the parabola shape shown for the gradient graph.
- Notice the graph is divided into sections using the stationary points:

| graph $y = f(x)$: | uphill | stationary | downhill |
|---|---|---|---|
| gradient graph $\frac{dy}{dx} = f'(x)$: | above $x$-axis | on the $x$-axis | below $x$-axis |

**Special cases:**

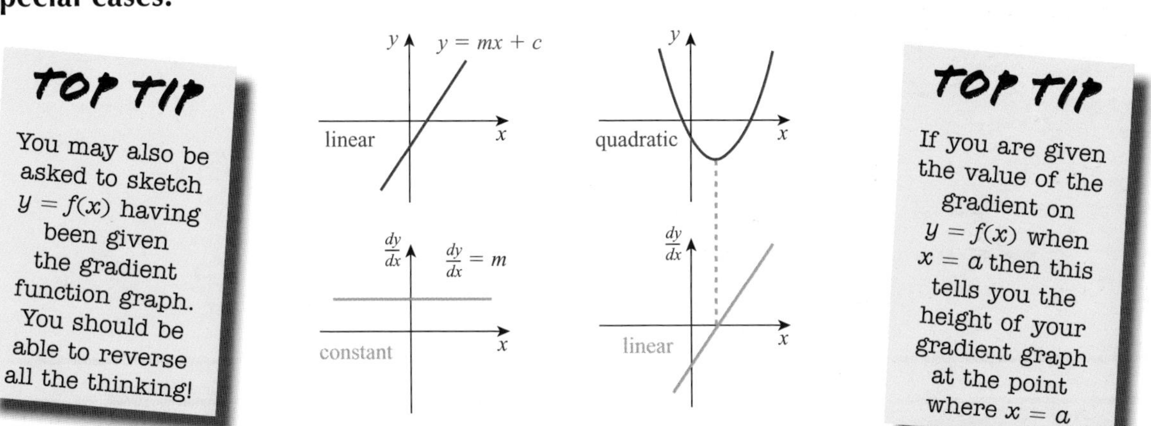


## Quick Test 23

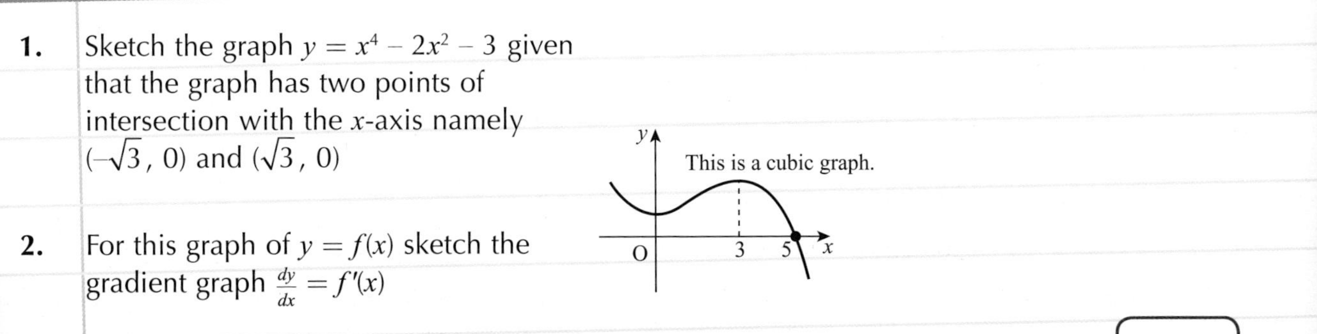

1. Sketch the graph $y = x^4 - 2x^2 - 3$ given that the graph has two points of intersection with the $x$-axis namely $(-\sqrt{3}, 0)$ and $(\sqrt{3}, 0)$
2. For this graph of $y = f(x)$ sketch the gradient graph $\frac{dy}{dx} = f'(x)$

# Further rules and techniques

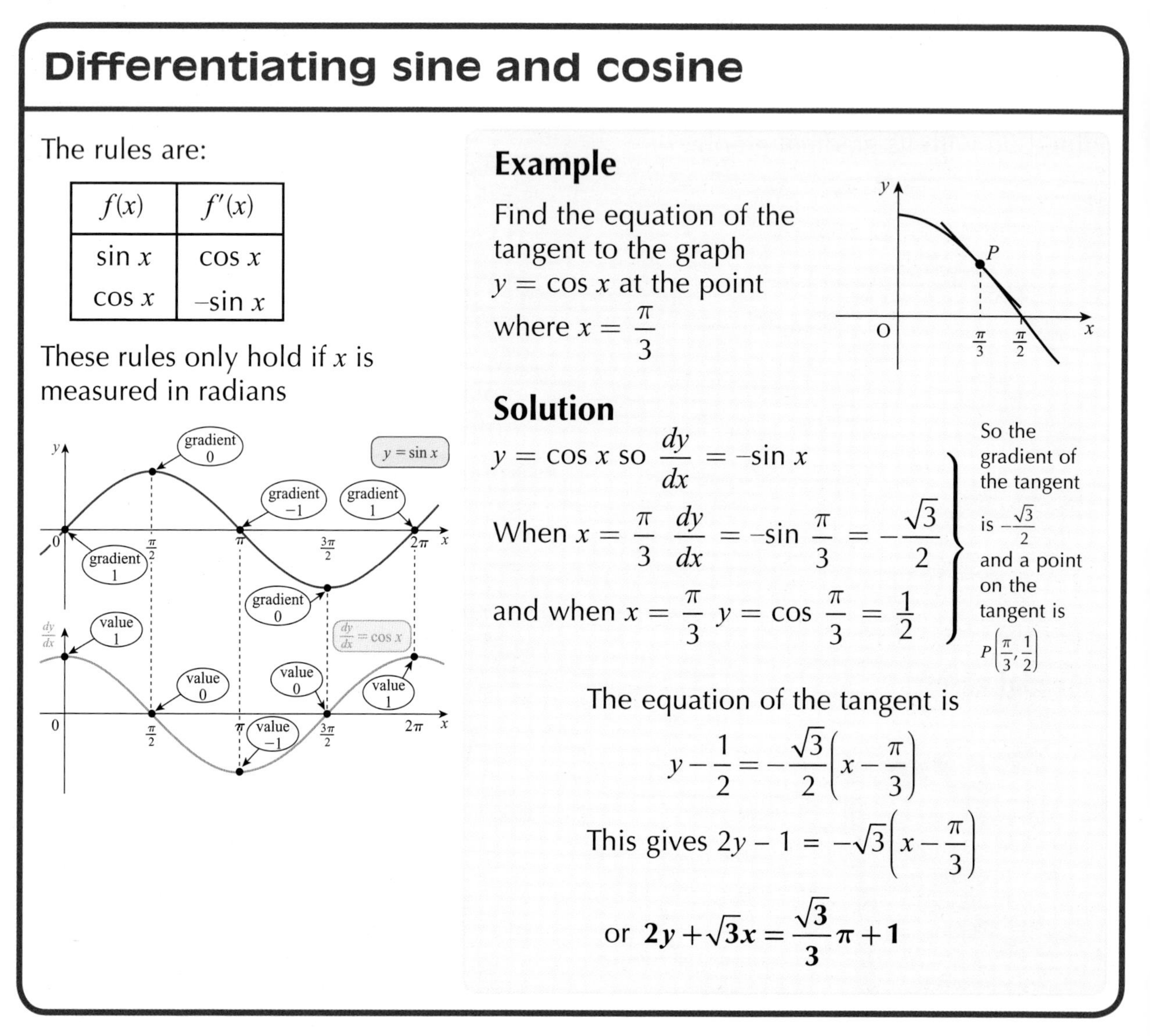

## Differentiating sine and cosine

The rules are:

| $f(x)$ | $f'(x)$ |
|---|---|
| $\sin x$ | $\cos x$ |
| $\cos x$ | $-\sin x$ |

These rules only hold if $x$ is measured in radians



**Example**

Find the equation of the tangent to the graph $y = \cos x$ at the point where $x = \frac{\pi}{3}$

**Solution**

$y = \cos x$ so $\frac{dy}{dx} = -\sin x$

When $x = \frac{\pi}{3}$ $\frac{dy}{dx} = -\sin\frac{\pi}{3} = -\frac{\sqrt{3}}{2}$

and when $x = \frac{\pi}{3}$ $y = \cos\frac{\pi}{3} = \frac{1}{2}$

So the gradient of the tangent is $-\frac{\sqrt{3}}{2}$ and a point on the tangent is $P\left(\frac{\pi}{3}, \frac{1}{2}\right)$

The equation of the tangent is

$$y - \frac{1}{2} = -\frac{\sqrt{3}}{2}\left(x - \frac{\pi}{3}\right)$$

This gives $2y - 1 = -\sqrt{3}\left(x - \frac{\pi}{3}\right)$

or $\mathbf{2y + \sqrt{3}x = \frac{\sqrt{3}}{3}\pi + 1}$

## The 'chain' rule

The 'chain' rule is used to extend the scope of the basic power rule and the two trig rules above:

**The power rule:** $y = x^n \Rightarrow \frac{dy}{dx} = nx^{n-1}$

**The extended power rule:** $y = (f(x))^n \Rightarrow \frac{dy}{dx} = n(f(x))^{n-1} \times f'(x)$

**An example:** $y = \sin^2 x = (\sin x)^2 \Rightarrow \frac{dy}{dx} = 2\,(\sin x)^1 \times \cos x = 2\sin x\cos x$

**A trig rule:** $y = \sin x \Rightarrow \frac{dy}{dx} = \cos x$

**The extended trig rule:** $y = \sin(f(x)) \Rightarrow \frac{dy}{dx} = \cos(f(x)) \times f'(x)$

**An example:** $y = \sin 3x \Rightarrow \frac{dy}{dx} = \cos 3x \times 3 = 3\cos x\ 3x$

**A trig rule:** $y = \cos x \Rightarrow \frac{dy}{dx} = -\sin x$

**The extended trig rule:** $y = \cos(f(x)) \Rightarrow \frac{dy}{dx} = -\sin(f(x)) \times f'(x)$

**An example:** $y = \cos(x^2) \Rightarrow \frac{dy}{dx} = -\sin(x^2) \times 2x = -2x\sin(x^2)$

## Working with the 'chain' rule

The general form of the 'chain' rule is: $y = g(h(x)) \Rightarrow \frac{dy}{dx} = g'(h(x)) \times h'(x)$

### Example 1

Differentiate a) $\cos(3x - \frac{\pi}{2})$ b) $\sqrt{x+1}$

**Solution**

a) $y = \cos(3x - \frac{\pi}{2}) \Rightarrow \frac{dy}{dx} = -\sin(3x - \frac{\pi}{2}) \times 3$

So $\frac{dy}{dx} = -3\sin(3x - \frac{\pi}{2})$

b) $y = \sqrt{x+1} = (x+1)^{\frac{1}{2}} \Rightarrow \frac{dy}{dx} = \frac{1}{2}(x+1)^{-\frac{1}{2}} \times 1$

So $\frac{dy}{dx} = \frac{1}{2(x+1)^{\frac{1}{2}}} = \frac{1}{2\sqrt{(x+1)}}$

### Example 2

Find $f'(\frac{3\pi}{4})$ where $f(x) = \frac{1}{3}\sin^3 x$

**Solution**

$f(x) = \frac{1}{3}(\sin x)^3$

$\Rightarrow f'(x) = \frac{1}{3} \times 3(\sin x)^2 \times \cos x$

So $f'\left(\frac{3\pi}{4}\right) = \sin^2 \frac{3\pi}{4} \cos \frac{3\pi}{4}$

$= \left(\frac{1}{\sqrt{2}}\right)^2 \times \left(-\frac{1}{\sqrt{2}}\right) = -\frac{1}{2\sqrt{2}}$

# Quick Test 24

1. Part of the graph $y = 2\sin^2 x$ is shown.
   Find the gradient of the curve at the point $P$ where $x = \frac{\pi}{4}$.
2. Differentiate a) $\sqrt{2 - 4x}$ b) $\frac{1}{\cos x}$
3. The function $g$ is defined by $g(x) = \frac{1}{x}$ where $x \neq 0$.
   Show that the graph $y = g(x)$ is always decreasing.

# Rates of change

## What is a rate of change?

Here is a distance/time graph:

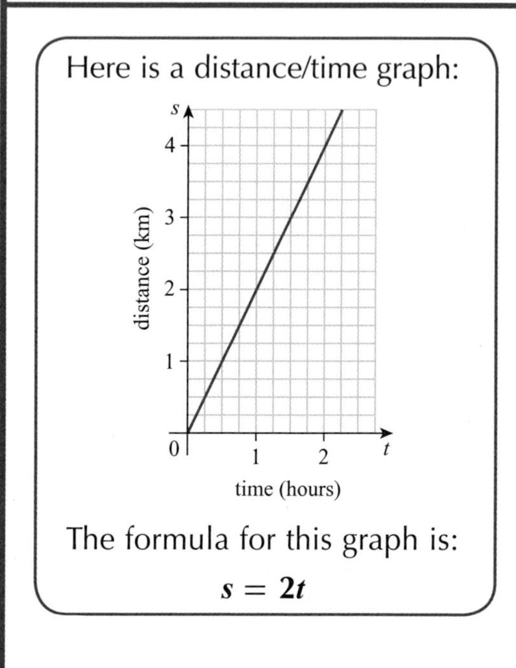


The formula for this graph is:

$$s = 2t$$

The distance changes at a rate of 2 km each hour.

You say that the rate of change of distance ($s$) with respect to time ($t$) is 2 km/hr.

In this case another name for this rate of change is speed.

The gradient of the graph measures the rate of change.

The speed is therefore given by: $\frac{ds}{dt} = 2$

In general for two variables $x$ and $y$ where the values of $y$ depend on the values of $x$ according to some rule $y = f(x)$ then the rate of change of $y$ with respect to $x$ is given by

$$\frac{dy}{dx} = f'(x)$$

and is shown by the gradient of the graph.

Notes:

- 'Displacement' and 'velocity' are used instead of 'distance' and 'speed' when direction matters e.g. height above or below a point, distance before or after a point etc. This would be indicated by a positive or negative sign.
- For a speed/time graph the rate at which speed changes with respect to time is acceleration.

**TOP TIP**

When asked to find the rate of change of $f(x)$ you need to differentiate.

## Working with the notation

Often the following letters are used:

$s$ for displacement

$v$ for velocity

$a$ for acceleration

$t$ for time

This gives:

$$\frac{ds}{dt} = v$$

The rate of change of displacement gives the velocity.

$$\frac{dv}{dt} = a$$

The rate of change of velocity gives the acceleration.

**Example**

An accelerating object has a displacement $s$ metres after $t$ seconds given by the formula: $\boldsymbol{s = t^2}$. What is its velocity after 1·5 seconds?

**Solution**

$s = t^2$

$v = \frac{ds}{dt} = 2t$

When $t = 1{\cdot}5$

$v = 2 \times 1{\cdot}5 = 3$

The velocity is 3 m/s (gradient at indicated point).

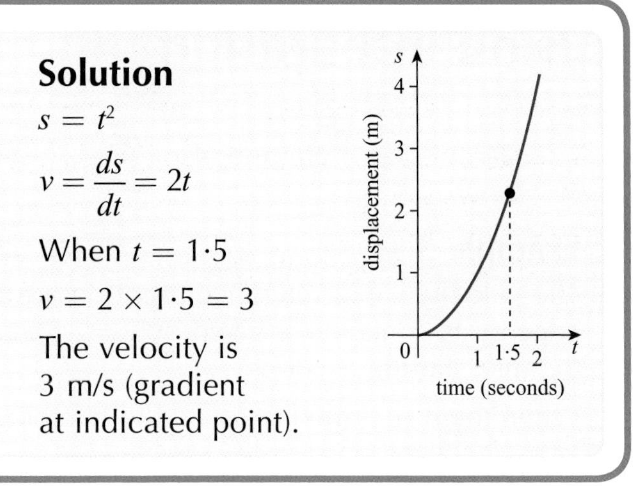


## Solving rate of change problems

**Example 1**

The displacement, $s$ cm, of a weight on a spring, $t$ seconds after release is given by $s = 50t - 100t^2$. Find its velocity when released and after $\frac{1}{4}$ second.

**Solution**

$$v = \frac{ds}{dt} = 50 - 200t$$

At time of release $t = 0$ so

$v = 50 - 200 \times 0 = \mathbf{50\ m/s}$

After $\frac{1}{4}$ second $t = \frac{1}{4}$ so

$$v = 50 - 200 \times \frac{1}{4} = \mathbf{0\ m/s}$$

(it has reached its greatest extent)

**Example 2**

The volume, $V$ cm$^3$, of a spherical balloon with radius $r$ cm is given by $V = \frac{4}{3}\pi r^3$. It is inflated.

Find the rate of change of $V$ with respect to $r$ when $r = 8$ cm.

**Solution**

$$V = \frac{4}{3}\pi r^3$$

so $\frac{dV}{dr} = 3 \times \frac{4}{3}\pi r^2 = 4\pi r^2$ This is the 'rate of change formula'.

When $r = 8$, $\frac{dV}{dr} = 4 \times \pi \times 8^2 \doteqdot 804$ (to 3 sig. figs).

This means that when the balloon has radius 8 cm and if the rate at which the volume is changing were to remain the same, then for an increase of 1 cm in the radius the volume would increase by 804 cm$^3$.

## Using integration

TOP TIP

The constant of integration is a crucial feature of these problems.

$$\frac{ds}{dt} = v \Rightarrow s = \int v\,dt \text{ and also } \frac{dv}{dt} = a \Rightarrow v = \int a\,dt$$

**Example 3**

The velocity $v$ of a reference point on a rotating disc when measured from the edge of its container is given by $v = -5\sin 2t$ where $t$ is the time in seconds from the disc starting to rotate.

When the disc started to rotate the reference point was $s = 5{\cdot}5$ cm from the edge.

How far is it from the edge after 3 seconds?

**Solution**

$v = \frac{ds}{dt} = -5\sin 2t \Rightarrow s = \int -5\sin 2t\,dt = \frac{5}{2}\cos 2t + C$

When $t = 0$, $s = 5{\cdot}5$ (the start of rotation)

so $5{\cdot}5 = \frac{5}{2}\cos 0 + C = \frac{5}{2} + C \Rightarrow C = 3$

so $s = \frac{5}{2}\cos 2t + 3$ (the displacement formula)

When $t = 3$ $s = \frac{5}{2}\cos 6 + 3 \doteqdot 5{\cdot}4$ cm

# Quick Test 25

1. The volume of a solid is given by $V(r) = \frac{2}{3}\pi r^3 + \pi r^2$.

   Find the rate of change of $V$ cm$^3$ with respect to the radius $r$ cm when $r = \frac{1}{2}$ cm.

2. A firework is launched. The height $h$ metres of the firework $t$ minutes after launch is given by the formula $h = 400t - 200t^2$.

   a) Find its speed at launch.

   b) Find its speed after 1 minute and explain your answer.

   c) Compare its speed after 2 minutes with its speed at launch and explain your result.

# Basic integration rules and techniques

## What is integration?

Integration reverses the process of Differentiation:

| | |
|---|---|
| $y = f(x)$ | $y = f(x) + C$ |
| $\Downarrow$ differentiation | $\Uparrow$ integration |
| $\frac{dy}{dx} = f'(x)$ | $\frac{dy}{dx} = f'(x)$ |

The integral sign $\int$ is used to indicate integration:

so $\int f'(x)\,dx = f(x) + C$

or $\int g(x)\,dx = G(x) + C$
where $G'(x) = g(x)$

$C$ is called the **Constant of Integration**.

Note: $\int f(x)\,dx$ is read 'the integral of $f(x)$ with respect to $x$' and is called an **Indefinite Integral**.

## Basic rules

| $f(x)$ | $\int f(x)\,dx$ |
|---|---|
| $x^n$ | $\frac{x^{n+1}}{n+1} + C \quad (n \neq -1)$ |
| $g(x) \pm h(x)$ | $\int g(x)\,dx \pm \int h(x)\,dx$ |
| (Integrate each term of a sum or difference.) | |
| $ag(x)$ | $a\int g(x)\,dx$ |
| (When a term is multiplied by a constant then integrate as normal and multiply the result by the same constant.) | |

Notes: Special cases

1. Integrating a constant: $\int k\,dx = kx + C$
2. $\int \frac{1}{x}\,dx = \int x^{-1}\,dx$.

   This integral does not follow the rule above. You do not need to know how to integrate $x^{-1}$.

**Example**

$\frac{dy}{dx} = 3x^2 - \frac{1}{x^2}$ Find $y$.

**Solution**

$$y = \int (3x^2 - x^{-2})\,dx$$

$$= \frac{3x^3}{3} - \frac{x^{-1}}{(-1)} + C$$

$$= \boldsymbol{x^3 + \frac{1}{x} + C}$$

**TOP TIP**

You should always check your integration answer by differentiating it. If it is correct you should get the expression you started with!

**Example**

Find $\int \left(\sqrt{x} - \frac{3}{\sqrt{x}}\right) dx$

**Solution**

First 'prepare' for integrating.

$$\int \left(x^{\frac{1}{2}} - \frac{3}{x^{\frac{1}{2}}}\right) dx = \int \left(x^{\frac{1}{2}} - 3x^{-\frac{1}{2}}\right) dx$$

3. A formula may involve a variable other than $x$, for example $f(t)$, and this may still be integrated:

   $\int f(t)\,dt$ (the integral with respect to $t$).

$$= \frac{x^{\frac{3}{2}}}{\frac{3}{2}} - \frac{3x^{\frac{1}{2}}}{\frac{1}{2}} + C$$

(Now double top and bottom of fractions)

$$= \frac{2x^{\frac{3}{2}}}{3} - 6x^{\frac{1}{2}} + C$$

## Integrating trig functions

The rules are:

| $f(x)$ | $\int f(x)\,dx$ |
|---|---|
| $\cos x$ | $\sin x + C$ |
| $\sin x$ | $-\cos x + C$ |

These rules only hold if $x$ is measured in radians.

**Example** Find $\int \left(\frac{3\cos x}{2} - \frac{2\sin x}{3}\right) dx$

**Solution**

$$\int \left(\frac{3\cos x}{2} - \frac{2\sin x}{3}\right) dx$$

$$= \int \left(\frac{3}{2}\cos x - \frac{2}{3}\sin x\right) dx$$

$$= \frac{3}{2}\sin x - \frac{2}{3}(-\cos x) + C$$

$$= \frac{3}{2}\sin x + \frac{2}{3}\cos x + C$$

## Solving differential equations

$\frac{dy}{dx} = f(x)$ is a **differential equation**. It has general solution $y = \int f(x)\,dx = F(x) + C$ where $F'(x) = f(x)$. If in addition a point $(a, b)$ is known on the graph $y = F(x) + C$ then the value of the constant $C$ can be found.

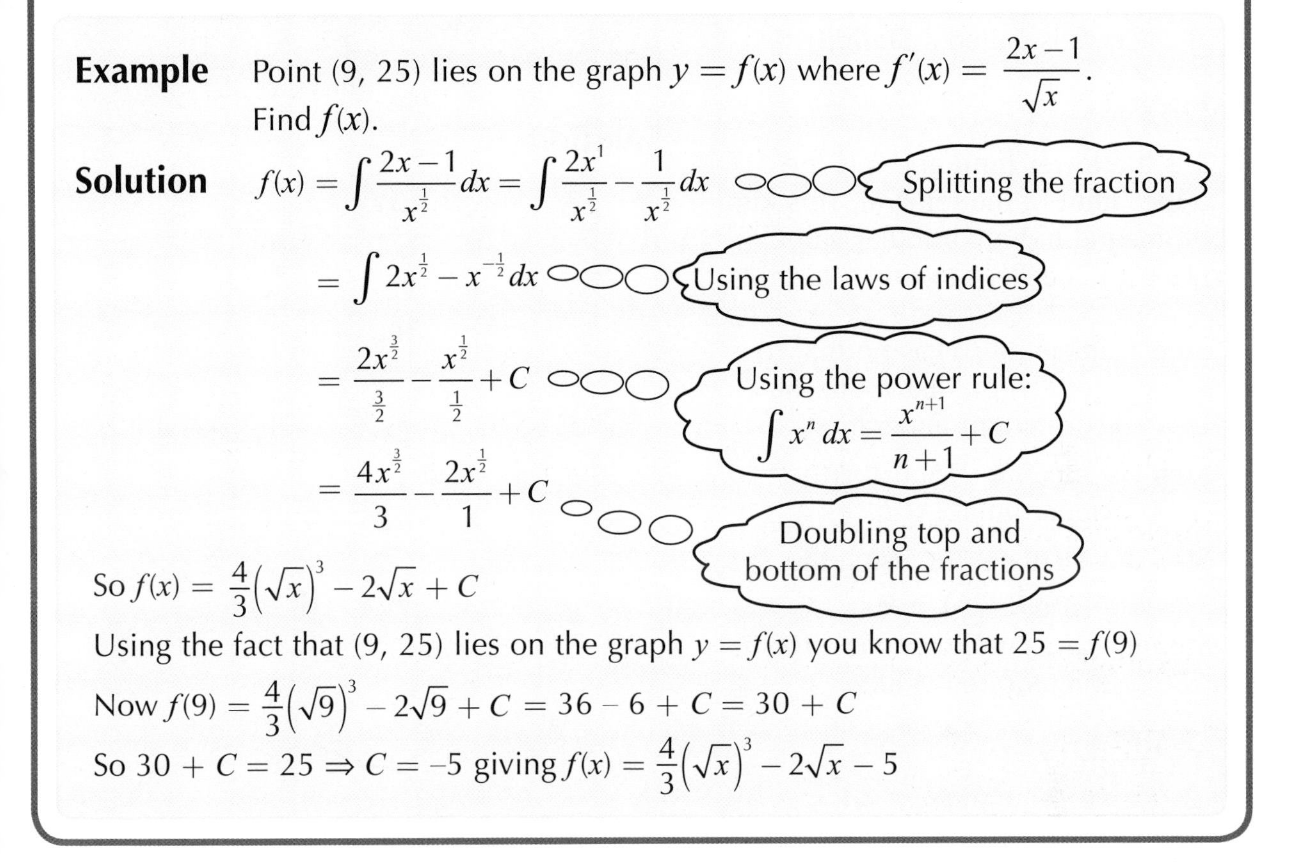

**Example** Point (9, 25) lies on the graph $y = f(x)$ where $f'(x) = \dfrac{2x-1}{\sqrt{x}}$. Find $f(x)$.

**Solution** $f(x) = \displaystyle\int \frac{2x-1}{x^{\frac{1}{2}}}dx = \int \frac{2x^1}{x^{\frac{1}{2}}} - \frac{1}{x^{\frac{1}{2}}}dx$

$= \displaystyle\int 2x^{\frac{1}{2}} - x^{-\frac{1}{2}}dx$

$= \dfrac{2x^{\frac{3}{2}}}{\frac{3}{2}} - \dfrac{x^{\frac{1}{2}}}{\frac{1}{2}} + C$

$= \dfrac{4x^{\frac{3}{2}}}{3} - \dfrac{2x^{\frac{1}{2}}}{1} + C$



So $f(x) = \frac{4}{3}\left(\sqrt{x}\right)^3 - 2\sqrt{x} + C$

Using the fact that (9, 25) lies on the graph $y = f(x)$ you know that $25 = f(9)$

Now $f(9) = \frac{4}{3}\left(\sqrt{9}\right)^3 - 2\sqrt{9} + C = 36 - 6 + C = 30 + C$

So $30 + C = 25 \Rightarrow C = -5$ giving $f(x) = \frac{4}{3}\left(\sqrt{x}\right)^3 - 2\sqrt{x} - 5$

# Quick Test 26

1. Find: a) $\displaystyle\int\left(\frac{1}{x^3} - 2x^2\right)dx$ b) $\displaystyle\int\left(5x^2 + 5 - \frac{3}{2x^2}\right)dx$ c) $\displaystyle\int\left(\frac{2}{3\sqrt{x}} + \sqrt{x}\right)dx$
2. Find: a) $\int(5x^3 + x + \sin x)\,dx$ b) $\int(4x - \cos x)\,dx$
3. If $y = \frac{2}{3}$ when $x = 1$ solve the differential equation $\dfrac{dy}{dx} = \dfrac{3-x^2}{3x^2}$ by finding $y$ in terms of $x$.

# Definite integrals and special integrals

## Definite integrals

$\int_a^b f(x)\,dx$ is called a **Definite Integral**.

$a$ is the **lower limit** and $b$ is the **upper limit** of the integral. This kind of integral has a particular value. Here are the steps to find this value:

**Step 1** Integrate $f(x)$ as normal to get $F(x)$ but leave out the constant of integration.

**Step 2** Calculate $F(b)$ using the upper limit $x = b$.

**Step 3** Calculate $F(a)$ using the lower limit $x = a$.

**Step 4** Calculate $F(b) - F(a)$. This is the required value.

### Example 1

Find the **exact** value of $\int_1^3 \left(\frac{3}{\sqrt{x}} - x\right) dx$

### Solution

$$\int_1^3 (3x^{-\frac{1}{2}} - x)\,dx = \left[\frac{3x^{\frac{1}{2}}}{\frac{1}{2}} - \frac{x^2}{2}\right]_1^3 = \left[6x^{\frac{1}{2}} - \frac{x^2}{2}\right]_1^3$$

$$= \left[6\sqrt{x} - \frac{x^2}{2}\right]_1^3$$

$$= \left(6\sqrt{3} - \frac{3^2}{2}\right) - \left(6\sqrt{1} - \frac{1^2}{2}\right)$$

$$= 6\sqrt{3} - \frac{9}{2} - 6 + \frac{1}{2}$$

$$= \mathbf{6\sqrt{3} - 10}$$

(Do not give a decimal approximation when an exact value is required.)

### Example 2

Evaluate $\int_{\frac{\pi}{6}}^{\frac{\pi}{3}} 2\sin x - 3\cos x\,dx$ giving your answer as a surd in its simplest form.

Remember the exact values diagram:



### Solution

$$\int_{\frac{\pi}{6}}^{\frac{\pi}{3}} 2\sin x - 3\cos x\,dx = [-2\cos x - 3\sin x]_{\frac{\pi}{6}}^{\frac{\pi}{3}}$$

$$= \left(-2\cos\tfrac{\pi}{3} - 3\sin\tfrac{\pi}{3}\right) - \left(-2\cos\tfrac{\pi}{6} - 3\sin\tfrac{\pi}{6}\right)$$

$$= \left(-2\times\frac{1}{2} - 3\times\frac{\sqrt{3}}{2}\right) - \left(-2\times\frac{\sqrt{3}}{2} - 3\times\frac{1}{2}\right)$$

$$= -1 - \frac{3\sqrt{3}}{2} + \sqrt{3} + \frac{3}{2}$$

$$= -\frac{2}{2} - \frac{3\sqrt{3}}{2} + \frac{2\sqrt{3}}{2} + \frac{3}{2}$$

$$= \frac{-2 - 3\sqrt{3} + 2\sqrt{3} + 3}{2} = \frac{1 - \sqrt{3}}{2}$$

**TOP TIP**

You should take great care with the negative signs in this sort of calculation, especially when subtracting the lower limit expression.

## Special integrals

It is not possible to extend the power rule and trig rules for integration in the same way that you did for differentiation using the 'chain rule'.

Here are the limited extended rules:

$$\int (ax+b)^n\,dx = \frac{(ax+b)^{n+1}}{a(n+1)} + C$$

$(n \neq -1,\ a \neq 0)$

$$\int \cos(ax+b)\,dx = \frac{\sin(ax+b)}{a} + C$$

$(a \neq 0)$

$$\int \sin(ax+b)\,dx = -\frac{\cos(ax+b)}{a} + C$$

$(a \neq 0)$

Only a linear expression like $ax + b$ is allowed and you divide by $a$, the coefficient of $x$.

**Example 1** Find $\int_{-1}^{1} \frac{4}{(5-3x)^2}\,dx$

**Solution**

$$\int_{-1}^{1} 4(5-3x)^{-2}\,dx = \left[\frac{4(5-3x)^{-1}}{-3\times(-1)}\right]_{-1}^{1}$$

$$= \left[\frac{4}{3(5-3x)}\right]_{-1}^{1} = \frac{4}{3(5-3\times 1)} - \frac{4}{3(5-3\times(-1))}$$

$$= \frac{4}{3\times 2} - \frac{4}{3\times 8} = \frac{4}{6} - \frac{1}{6} = \frac{3}{6} = \mathbf{\frac{1}{2}}$$

**Example 2**

Find a) $\int \sin 3x\,dx$ b) $\int 4\cos(2x - \frac{\pi}{6})\,dx$

**Solution**

a) $\int \sin 3x\,dx = \frac{-\cos 3x}{3} + C = -\frac{1}{3}\cos 3x + C$

b) $\int 4\cos(2x - \frac{\pi}{6})\,dx = \frac{4\sin(2x - \frac{\pi}{6})}{2} + C$

$= 2\sin(2x - \frac{\pi}{6}) + C$

**Example 3**

a) Use $\cos 2x = 2\cos^2 x - 1$ to write $\cos^2 x$ in terms of $\cos 2x$

b) Hence find $\int \cos^2 x\,dx$

**Solution**

a) Rearranging gives

$$\cos^2 x = \frac{1}{2}\cos 2x + \frac{1}{2}$$

b) $\int \cos^2 x\,dx = \int \frac{1}{2}\cos 2x + \frac{1}{2}\,dx$

$= \frac{\sin 2x}{2\times 2} + \frac{1}{2}x + C$

$= \frac{1}{4}\sin 2x + \frac{1}{2}x + C$

**TOP TIP**

You are given this table of standard integrals in your exam:

| $f(x)$ | $\int f(x)x\,dx$ |
|---|---|
| $\sin ax$ | $-\frac{1}{a}\cos ax + C$ |
| $\cos ax$ | $\frac{1}{a}\sin ax + C$ |

## Quick Test 27

1. Find the exact value of: $\int_1^4 \left(\sqrt{x} + \frac{1}{\sqrt{x}}\right)dx$

2. Find the exact value of $\int_0^{\frac{\pi}{4}} \frac{1}{2}\cos x - \sin x\,dx$

3. a) Use $\cos 2x = 1 - 2\sin^2 x$ to write $\sin^2 x$ in terms of $\cos 2x$

   b) Hence find $\int \sin^2 x\,dx$

   c) Evaluate $\int_0^{\pi} \sin^2 x\,dx$

# Sample Unit 2 test questions

## 1.1 Applying algebraic skills to solve equations

1. $g(x)$ is a cubic function. Here are some facts concerning $g(x)$:
   - The graph $y = g(x)$ crosses the $x$-axis at $(-2, 0)$
   - $x + 3$ is a factor of $g(x)$
   - If $g(x)$ is divided by $x - 1$ the remainder is zero

   Find the roots of the equation $g(x) = 0$
2. Myla draws the graph of $y = 2x^2 - x + a$ using an App on her phone.

   She notices that it has no $x$-axis intercepts.

   What is the range of values for $a$ in this case?
3. A cubic function $f$ is defined by $f(x) = x^3 - x^2 - 16x - 20$

   a) Factorise $f(x)$ fully. b) Solve $f(x) = 0$

## 1.2 Applying trig skills to solve equations

1. Solve $2\cos 3x° = 1$ for $0 < x < 120$
2. Solve $\cos\theta° - 5\sin 2\theta° = 0$ for $0 \leq \theta \leq 90$
3. Carla was asked to write $4\sin\theta° - 3\cos\theta°$ in the form $k\sin(\theta - \alpha)°$

   She correctly calculated $k = 5$ and $\alpha = 36{\cdot}9$

   Hence she wrote $4\sin\theta° - 3\cos\theta° = 5\sin(\theta - 36{\cdot}9)°$

   Use her result to solve $4\sin 2\theta° - 3\cos 2\theta° = 2{\cdot}75$ for $0 \leq \theta \leq 90$

## 1.3 Applying calculus skills of differentiation

1. Differentiate $f(x)$ where $f(x) = -2\cos x$
2. Find $\frac{dy}{dx}$ where $y = \frac{5\sqrt{x}}{x} - 2\sqrt{x}$ $(x > 0)$
3. Part of the graph $y = \frac{1}{2}x^2 + 3x - 2$ is shown in the diagram.

   Find the equation of the tangent to the curve at the point $P$ where $x = 2$

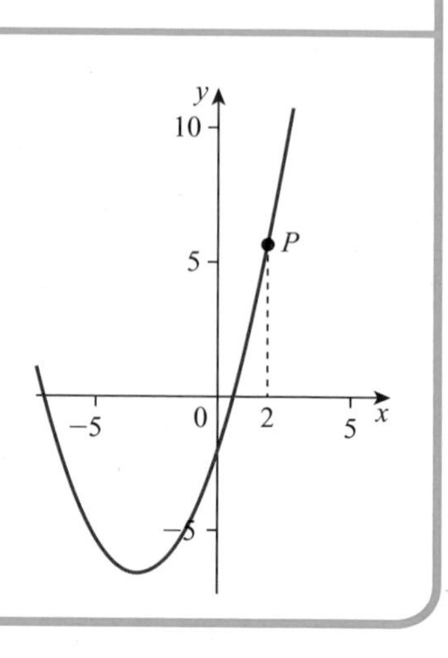

4. A stone is thrown into the air.

   The height $h$ metres $t$ seconds after it was thrown is given by the formula:

$$h = 16t - 4t^2$$

   a) Find the velocity of the stone when it was thrown.

   b) Find the velocity of the stone after 2 seconds and explain your answer.

## 1.4 Applying calculus skills of integration

1. Find $\int \frac{1}{2}\cos x\,dx$
2. For the function $g$ it is known that $g'(x) = (5 + x)^{-2}$

   Find $g(x)$
3. Find $\int \frac{1}{x^{\frac{1}{2}}} - 2x^{\frac{1}{2}}\,dx$ where $x > 0$
4. Evaluate the definite integral $\int_{-2}^{2} (x + 2)^3\,dx$

# Sample end-of-course exam questions (Unit 2)

## Non-calculator

1. The functions $f$, $g$ and $h$ are defined on the set of Real numbers by:

   $f(x) = 2x^3 + x - 3$

   $g(x) = -3x^2 + x - 2$

   $h(x) = 2x^3 + 3x^2 - 1$

   a) i) Show that $x + 1$ is a factor of $h(x)$.

   ii) Hence factorise $h(x)$ fully.

   iii) Solve $h(x) = 0$.

   b) The two curves $y = f(x)$ and $y = g(x)$ share a common tangent at point $T$.

   Find the coordinates of $T$.

2. Find the value of $\int_{-1}^{1} \frac{6x^3 - x}{3x^3}\, dx$.

3. Find $\frac{dy}{dx}$ given that $y = \frac{1}{\sin^2 x}$.

4. Differentiate $\frac{2x+6}{\sqrt{x}}$ with respect to $x$.

5. The diagram shows a sketch of part of a cubic graph $y = f(x)$ with stationary points $(0, a)$ and $(b, c)$.

   Sketch the graph of $y = f'(x)$.

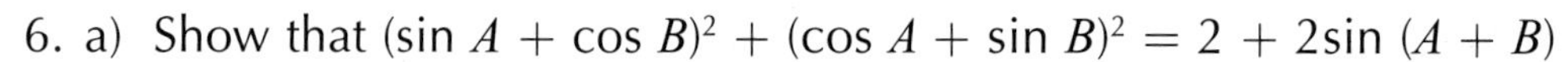

6. a) Show that $(\sin A + \cos B)^2 + (\cos A + \sin B)^2 = 2 + 2\sin(A + B)$

   b) Hence, if $(\sin A + \cos B)^2 + (\cos A + \sin B)^2 = 3$, find two possible values for angle $(A + B)$ between 0 and $2\pi$.

7. Part of the graphs of $y = 2\sqrt{x+1}$ and $2y - 3x = 6$ are shown in the diagram. A tangent to the curve is drawn parallel to the given straight line.

   Find the $x$-coordinate of the point of contact of this tangent to the curve.

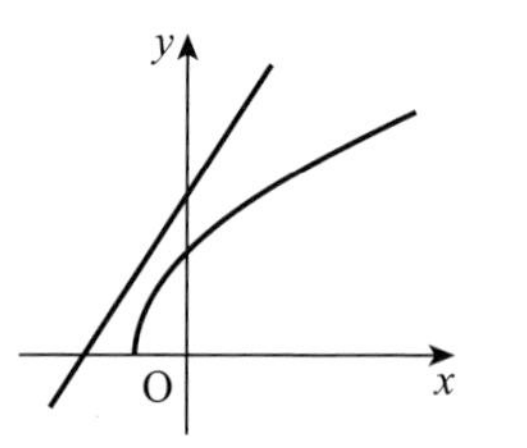

## Calculator allowed

1. $f(x) = \cos x° - 5\sin x°$
   a) Express $f(x)$ in the form $k\cos(x + \alpha)°$ where $k > 0$ and $0 \leq \alpha \leq 360$.
   b) Hence solve $f(x) = 1$ for $0 \leq x < 360$.
   c) The graph $y = f(x)$ cuts the $x$-axis at the point $(a, 0)$ where $180 < a < 270$. Find the value of $a$.
2. a) Show that $\cos^2 x° - \cos 2x° = 1 - \cos^2 x°$.
   b) Hence solve the equation $3\cos^2 x° - 3\cos 2x° = 8\cos x°$ in the interval $0 \leq x < 90$.
3. The roots of the equation $(kx + 2)(x + 3) = 8$ are equal. Find the values of $k$.
4. A preliminary partial sketch of the curve with equation $y = 3 + 2x^2 - x^4$ is shown in the diagram.
   a) Find the coordinates of the stationary points on the curve.
   b) Confirm the information in the sketch by determining the nature of the stationary points.
5. The function $f$ is defined by $f(\theta) = 3\cos\theta° - \sin\theta°$.
   a) Show that $f(\theta)$ can be expressed in the form $f(\theta) = k\cos(\theta + \alpha)°$ where $k > 0$ and $0 \leq \alpha \leq 360$ and determine the values of $k$ and $\alpha$.
   b) Hence find the maximum and minimum values of $f(\theta)$ and the corresponding values of $\theta$, where $\theta$ lies in the interval $0 \leq \theta < 360$.
   c) Write down the minimum value of $\sqrt{10}(3\cos\theta° - \sin\theta°) + 10$.
6. The function $f(x) = x^4 + 4x^3 + 5x^2 + 14x + 24$
   $= (x + a)(x + b)(x^2 - x + 4)$
   a) Find the values of $a$ and $b$.
   b) Hence show that the equation $f(x) = 0$ has only two real solutions.
7. Given that $5 \times 3^\alpha = 2$ where $\alpha = \cos^2 x - \sin^2 x$, calculate the smallest possible positive value of $x$.

# Gradient and the straight line revisited

## What is gradient?

Gradient is a **number** that measures the slope of a line. Divide the **vertical distance** by the **horizontal distance**.

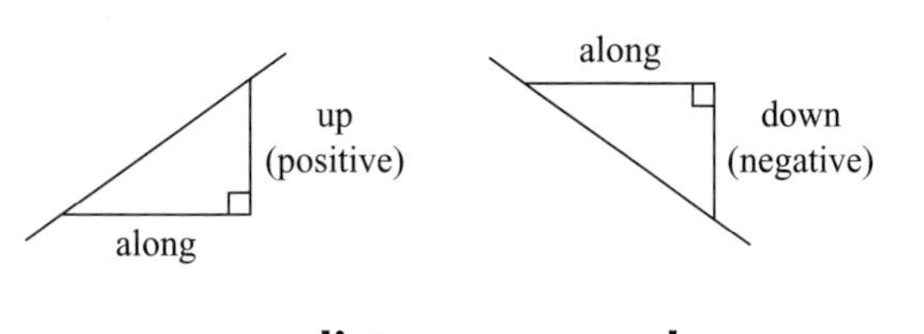


$$\textbf{gradient} = \frac{\textbf{distance up or down}}{\textbf{distance along}}$$

Here is some notation to do with gradient:

- $m$ unknown value of a gradient
- $m_{AB}$ the gradient of the line $AB$
- $m_{\perp}$ the gradient of a perpendicular line
- $m_1$, $m_2$ the values of two different gradients

## The gradient formula

If you know the two pairs of coordinates for two points $A$ and $B$ here is how to find $m_{AB}$:

$B(x_2, y_2)$

$A(x_1, y_1)$

$$m_{AB} = \frac{y_2 - y_1}{x_2 - x_1}$$

$y$-coordinate difference

$x$-coordinate difference

Notes:

1. $x_1 \neq x_2$. If $x_1 = x_2$ you get a 'vertical' line which has **no** gradient.
2. $\frac{y_1 - y_2}{x_1 - x_2}$ gives the same result.

   you can swap **both** top and bottom order but not just one.

**Example** Find the gradient of $AB$ where $A$ and $B$ have coordinates (–1, 1) and (5, –2).

**Solution** $m_{AB} = \dfrac{1-(-2)}{-1-5} = \dfrac{3}{-6} = -\dfrac{1}{2}$ Note that $\dfrac{-2-1}{5-(-1)} = \dfrac{-3}{6} = -\dfrac{1}{2}$ gives the same result.

## The equation of a straight line I

Here is a typical equation of a straight line:

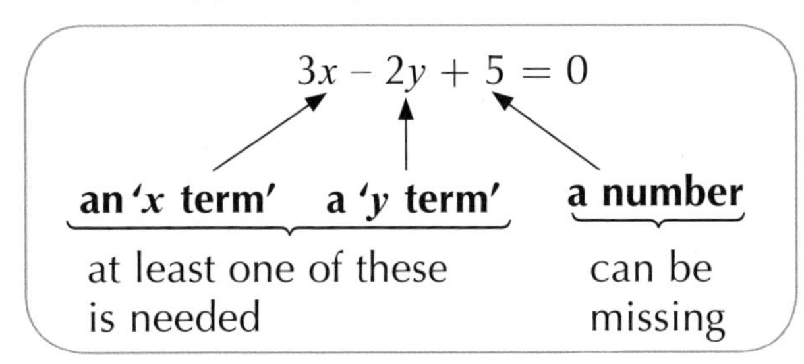


Unless the line is parallel to the $y$-axis it is always possible to rearrange the equation into the form:

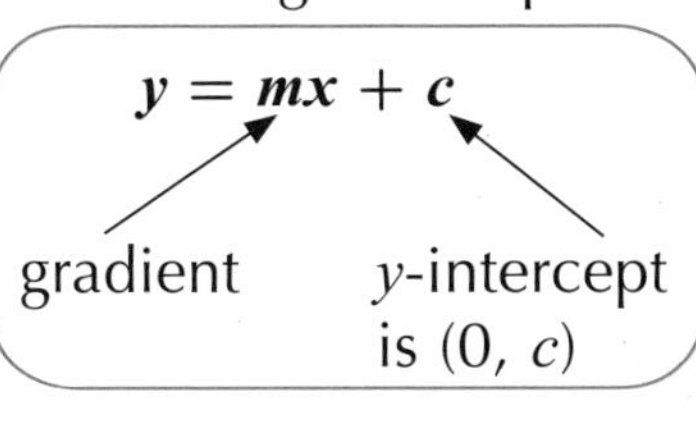


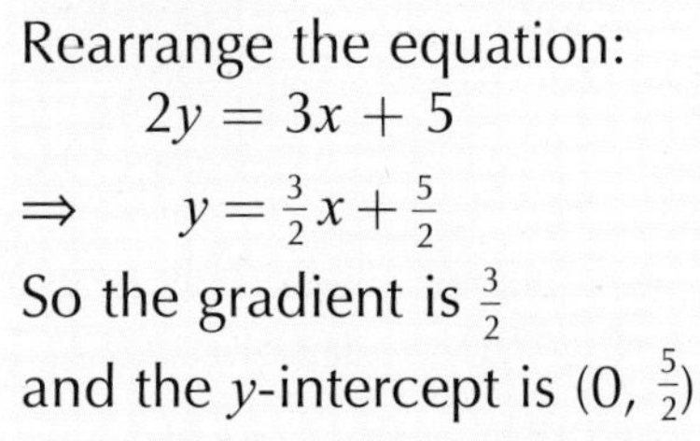

**Example**

Find the gradient and $y$-intercept of the line with equation $3x - 2y + 5 = 0$

**Solution**

Rearrange the equation:

$$2y = 3x + 5$$

$$\Rightarrow \quad y = \tfrac{3}{2}x + \tfrac{5}{2}$$

So the gradient is $\frac{3}{2}$ and the $y$-intercept is $(0, \frac{5}{2})$

## Special cases

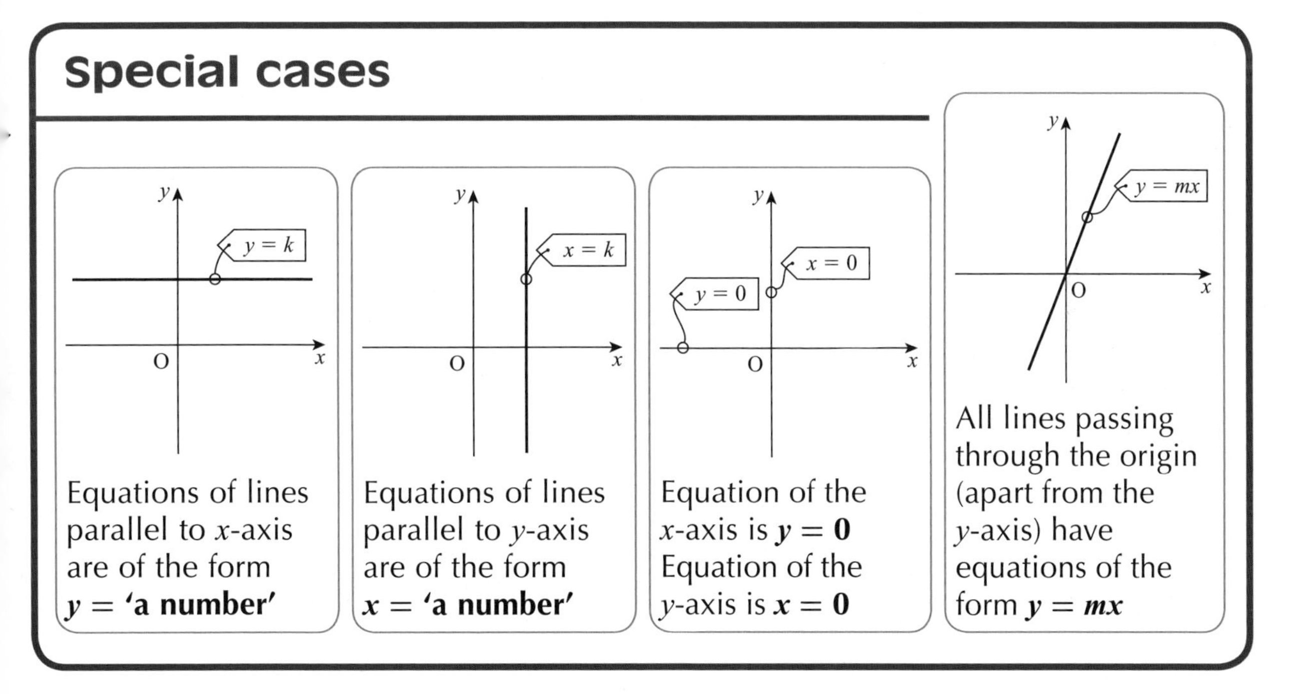


Equations of lines parallel to $x$-axis are of the form $y$ = **'a number'**

Equations of lines parallel to $y$-axis are of the form $x$ = **'a number'**

Equation of the $x$-axis is $\boldsymbol{y = 0}$
Equation of the $y$-axis is $\boldsymbol{x = 0}$

All lines passing through the origin (apart from the $y$-axis) have equations of the form $\boldsymbol{y = mx}$

# Quick Test 28

1. Find the gradient $m_{PQ}$ where $P(-1, 4)$ and $Q(-4, 3)$
2. a) Find the gradient and $y$-intercept of the line with equation $3x - 2y = 4$
   b) Find the equation of the straight line with the same $y$-axis intercept but with gradient $\frac{2}{3}$
3. Find the equation of the line through $(-2, -1)$ and $(-2, 7)$.

# Working with the gradient

## Gradient and angles

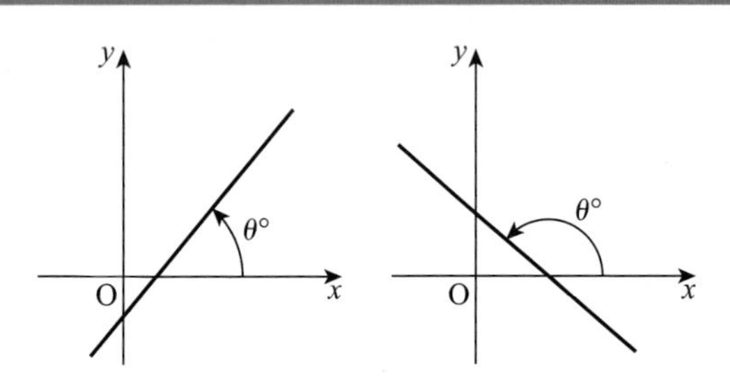


tan $\theta°$ is positive for $0 < \theta < 90$

tan $\theta°$ is negative for $90 < \theta < 180$

Assuming the scales are the same on both axes then...

**tan $\theta$ = the gradient of the line ($m$)**

### Example

Find the angle that a line with gradient $-\frac{1}{3}$ makes with the positive direction of the $x$-axis.

### Solution

Suppose the angle is $\theta°$, then...

$$\tan\theta° = -\tfrac{1}{3}$$

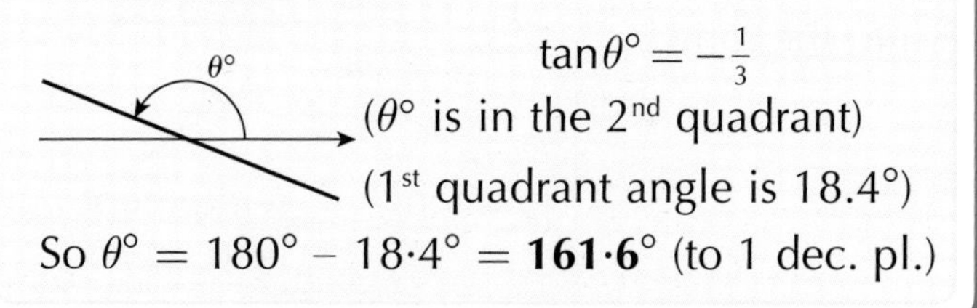


($\theta°$ is in the 2nd quadrant)

(1st quadrant angle is 18.4°)

So $\theta° = 180° - 18{\cdot}4° =$ **161·6°** (to 1 dec. pl.)

## Parallel lines and collinear points

**$m_1 = m_2$**

If one is true ↕ then so is the other

**The lines are parallel**

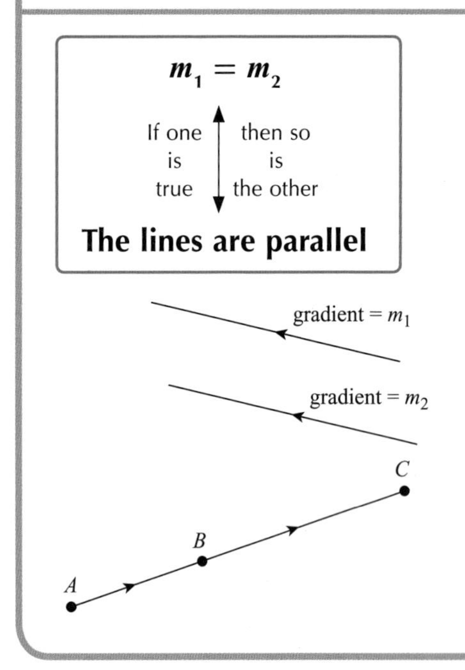


### Example

Show that the points $A(-1, -5)$, $B(1, -4)$ and $C(7, -1)$ are collinear.

### Solution

$$m_{AB} = \frac{-4-(-5)}{1-(-1)} = \frac{1}{2} \text{ and}$$

$$m_{BC} = \frac{-1-(-4)}{7-1} = \frac{1}{2}$$

So $AB$ and $BC$ are parallel.

Since they share a common point $B$ they are collinear.

## Perpendicular lines

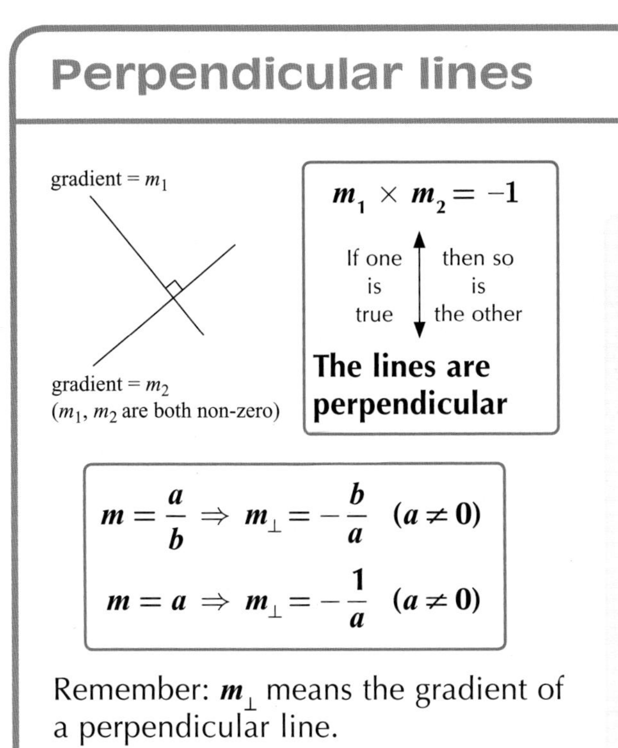


$m_1 \times m_2 = -1$

If one is true ⇕ then so is the other

**The lines are perpendicular**

$$m = \frac{a}{b} \Rightarrow m_\perp = -\frac{b}{a} \quad (a \neq 0)$$

$$m = a \Rightarrow m_\perp = -\frac{1}{a} \quad (a \neq 0)$$

Remember: $m_\perp$ means the gradient of a perpendicular line.

**TOP TIP**

To find a perpendicular gradient you: "change the sign and invert"

$m = \frac{2}{3}$ becomes $m_\perp = -\frac{3}{2}$

**Example**

A triangle $ABC$ has vertices $A(-3, -1)$, $B(-1, 2)$ and $C(5, -2)$. Show that it is right-angled.

**Solution**

$$m_{AB} = \frac{-1-2}{-3-(-1)} = \frac{-3}{-2} = \frac{3}{2}$$

$$m_{BC} = \frac{2-(-2)}{-1-5} = \frac{4}{-6} = -\frac{2}{3}$$

Since $m_{AB} \times m_{BC} = \frac{3}{2} \times \left(-\frac{2}{3}\right) = -1$

then $AB \perp BC$ ($AB$ is perpendicular to $BC$) and so $\Delta ABC$ is right-angled at $B$.

## The equation of a straight line II

If you know:

**Fact 1** the gradient: $m$

**Fact 2** a point on the line: $(a, b)$

then the equation of the straight line is:

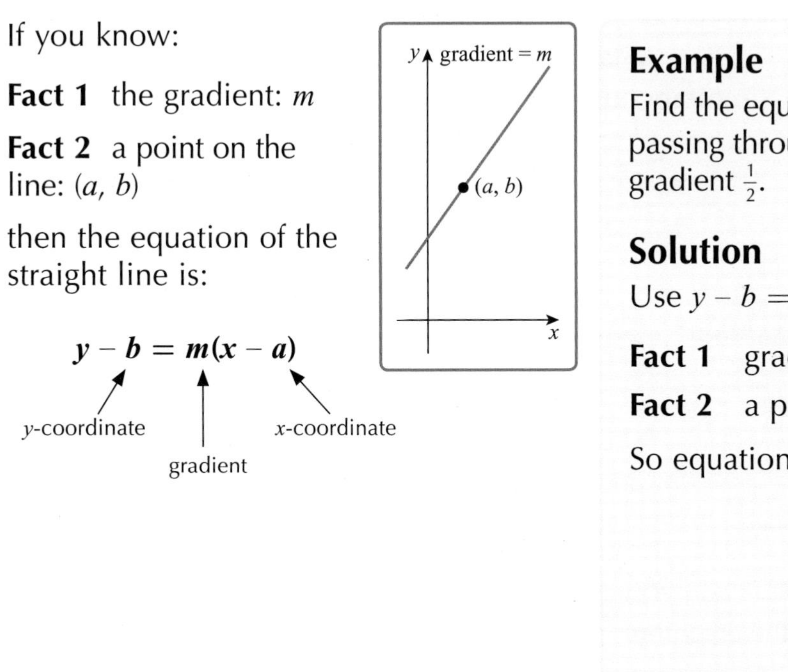


$$y - b = m(x - a)$$

$y$-coordinate, gradient, $x$-coordinate

**Example**

Find the equation of the line passing through (2, –3) with gradient $\frac{1}{2}$.

**Solution**

Use $y - b = m(x - a)$

**Fact 1** gradient: $\frac{1}{2}$

**Fact 2** a point on line: (2, –3)

So equation is $y - (-3) = \frac{1}{2}(x - 2)$

$$y + 3 = \frac{1}{2}(x - 2)$$

$$2y + 6 = x - 2$$

$$\mathbf{2y - x = -8}$$

Get rid of this fraction by doubling

**Example**

Find the equation of the line passing through $A(-1, 3)$ and $B(4, -2)$.

**Solution**

Find the gradient:

$$m_{AB} = \frac{3-(-2)}{-1-4} = \frac{5}{-5} = -1$$

Use $y - b = m(x - a)$

**Fact 1** gradient: $-1$

**Fact 2** a point on line: $(-1, 3)$

So equation is $y - 3 = 1(x - (-1))$

$\Rightarrow \quad y - 3 = -(x + 1)$

$\Rightarrow \quad y - 3 = -x - 1$

giving $\boldsymbol{y + x = 2}$

## Points of intersection

To find where a line crosses the axes:

|  | | |
|---|---|---|
| to find: | ***x*-axis intercept** ⟷ | **Set $y = 0$ in the equation** |
| to find: | ***y*-axis intercept** ⟷ | **Set $x = 0$ in the equation** |

To find where two lines intersect:

**equation of 1st line** → **solve these simultaneously**
**equation of 2nd line** → 

**Example**

Find the point of intersection of the lines $3y = 2x + 4$ and $3x = 7 - 2y$

**Solution**

after rearranging equations

$$\left.\begin{aligned} 3y - 2x &= 4 \\ 2y + 3x &= 7 \end{aligned}\right\} \begin{aligned} &\times 3 \rightarrow 9y - 6x = 12 \\ &\times 2 \rightarrow 4y + 6x = 14 \end{aligned}$$

Add: $13y = 26$

$y = 2$

Put $y = 2$ in $2y + 3x = 7 \Rightarrow 4 + 3x = 7$

So $3x = 3$ giving $x = 1$

**(1, 2)** is the point of intersection.

**TOP TIP**

To prove perpendicularity: Use $m_1$ and $m_2$ for the two gradients and then calculate $m_1 \times m_2$ showing it equals $-1$

## Quick Test 29

1. For the line $AB$ where $A(-3, -5)$ and $B(5, -1)$ find
   a) the gradient   b) the equation
   c) the angle it makes with the positive direction of the $x$-axis
2. Show that the lines $y + 2x + 1 = 0$ and $2y - x - 3 = 0$ are perpendicular and find their point of intersection.

# Problem solving using the gradient

## The distance formula

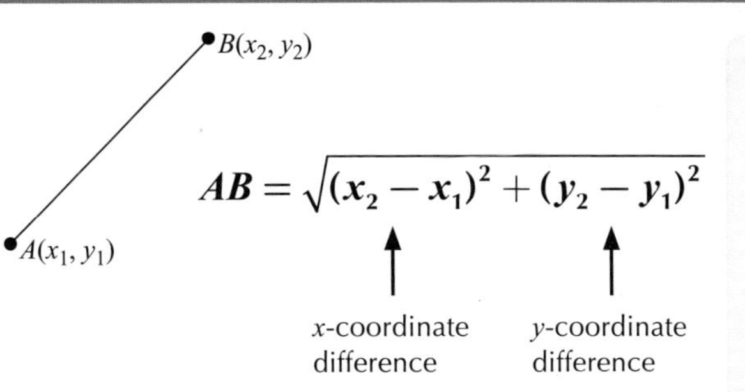


Note:

$AB^2 = (x_2 - x_1)^2 + (y_2 - y_1)^2$

if you wish to avoid the square root.

### Example

Show that triangle $ABC$ with vertices $A(1, 2)$, $B(3, 0)$ and $C(-1, -2)$ is isosceles.

### Solution

$$AC = \sqrt{(1-(-1))^2 + (2-(-2))^2} = \sqrt{2^2 + 4^2}$$
$$= \sqrt{4+16} = \sqrt{20}$$

$$BC = \sqrt{(3-(-1))^2 + (0-(-2))^2} = \sqrt{4^2 + 2^2}$$
$$= \sqrt{16+4} = \sqrt{20}$$

So $AC = BC$ and the triangle is isosceles.

## The midpoint formula

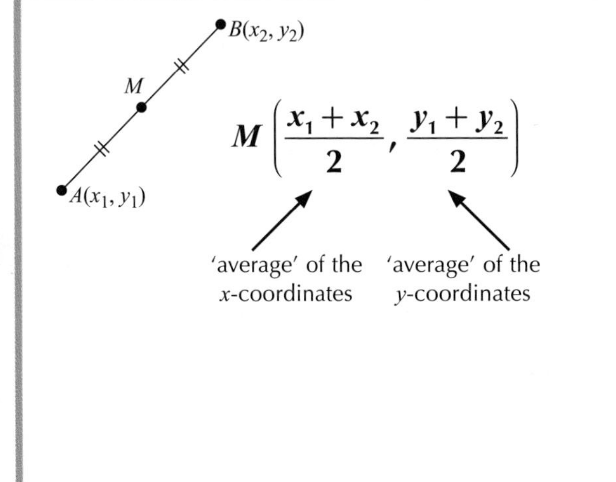


### Example

Find the coordinates of $M$, the midpoint of $CD$, where $C$ is the point $(-1, 5)$ and $D$ is $(-5, 2)$.

### Solution

$$M\left(\frac{-1+(-5)}{2}, \frac{5+2}{2}\right)$$

$$= M\left(\frac{-6}{2}, \frac{7}{2}\right) = M\left(-3, \frac{7}{2}\right)$$

## Perpendicular bisectors

The **perpendicular bisector** of line $AB$ is a line with these two properties:

**Property 1** It is perpendicular to $AB$

**Property 2** It passes through the midpoint of $AB$

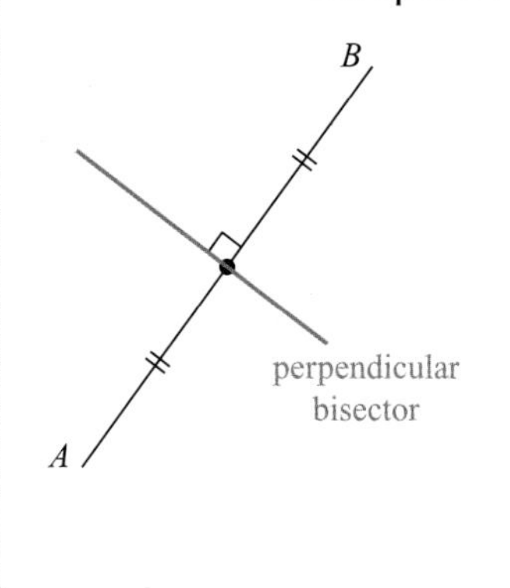


**Example**

Find the equation of the perpendicular bisector of the line joining $P(2, 3)$ and $Q(10, 1)$

**Solution**

The midpoint of $PQ$ is $M\left(\frac{2+10}{2}, \frac{3+1}{2}\right) = M(6, 2)$

also $m_{PQ} = \frac{1-3}{10-2} = \frac{-2}{8} = -\frac{1}{4}$

$\Rightarrow m_{\perp} = 4$ (perpendicular gradient)

Using '$y - b = m(x - a)$' the equation is:

$y - 2 = 4(x - 6)$

$\Rightarrow y - 2 = 4x - 24$

giving $y - 4x = -22$

## Special lines in triangles

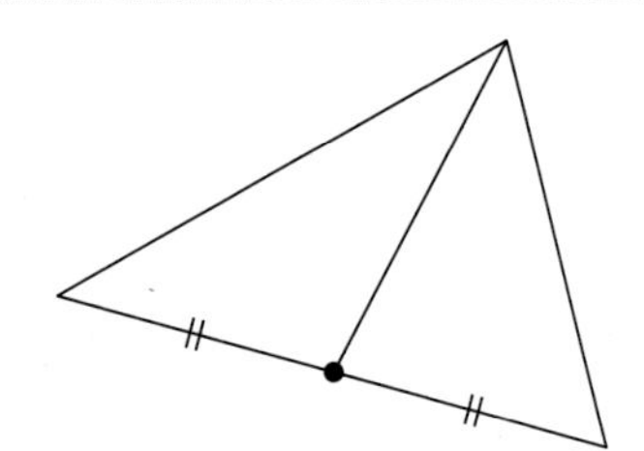

A **median** joins a vertex to the midpoint of the opposite side.

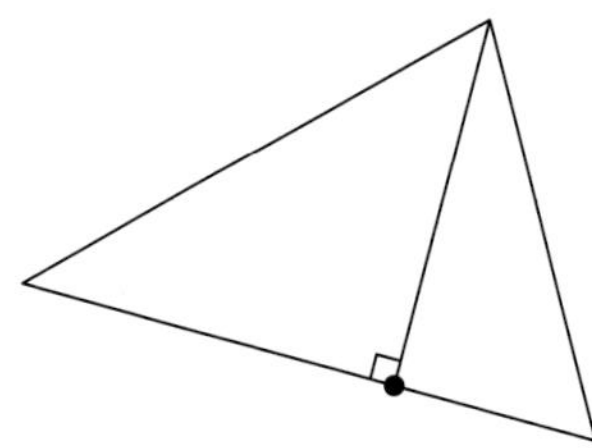

An **altitude** is a line through a vertex perpendicular to the opposite side.

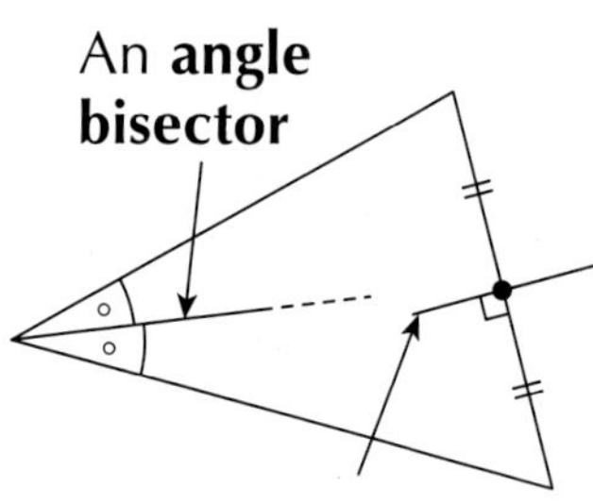


A **perpendicular bisector** of a side.

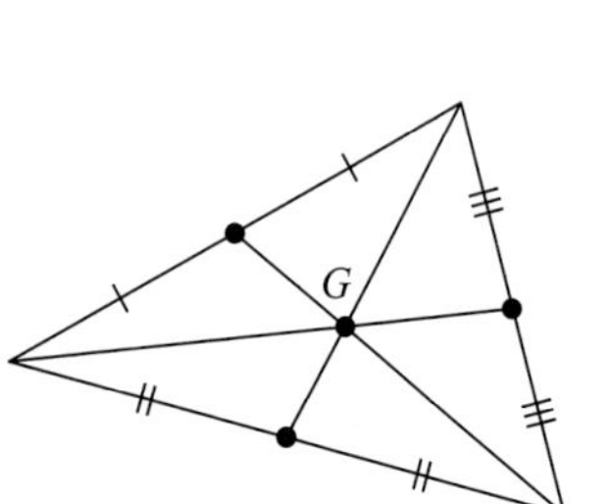


The three medians are **concurrent**. They meet at $G$, the **centroid** of the triangle.

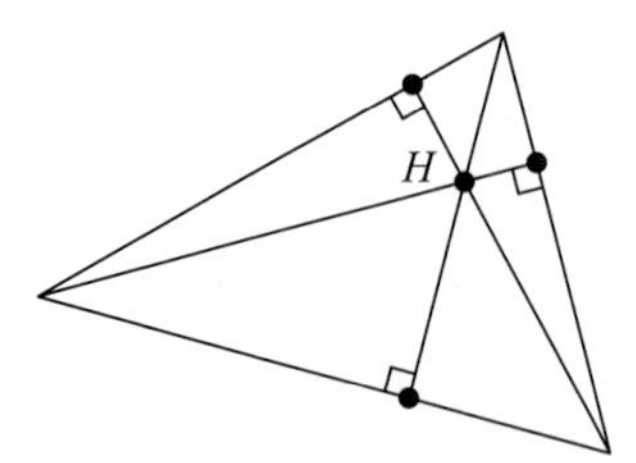


The three altitudes are also concurrent (meeting at $H$, the **orthocentre**).

Notes:
1. The centroid $G$ divides each median in the ratio 2:1 (vertex to midpoint).
2. 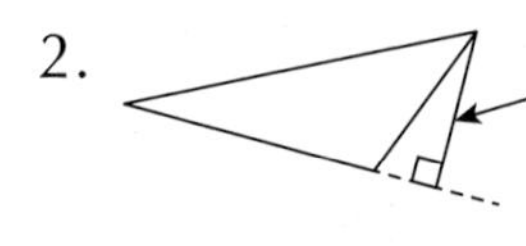 Altitudes can lie 'outside' the triangle.
3. The three angle bisectors are concurrent as are the three perpendicular bisectors of the sides.

**TOP TIP**

Lines that are **concurrent** meet in one point called the point of concurrency.

### Example

A triangle has vertices $P(-2, 3)$, $Q(6, -1)$ and $R(-4, -5)$. Find the coordinates of $G$, the point of intersection of the medians $PS$ and $RT$.

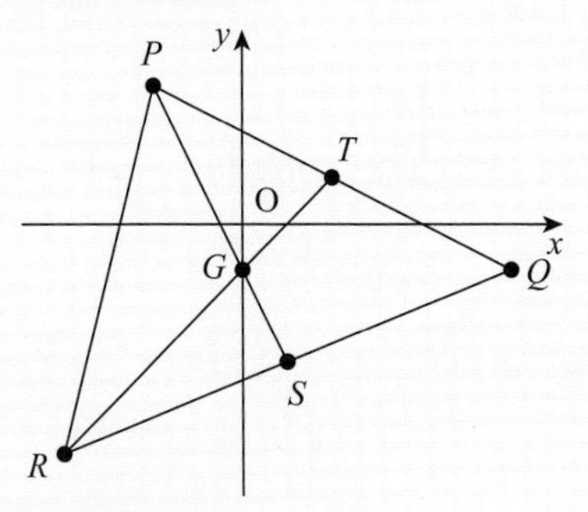


### Solution

$S\left(\frac{-4+6}{2}, \frac{-5+(-1)}{2}\right) = S(1, -3)$

also $m_{PS} = \left(\frac{-3-3}{1-(-2)}\right) = \frac{-6}{3} = -2$

Equation of $PS$: $y - (-3) = -2(x - 1)$

$\Rightarrow y + 3 = -2x + 2 \Rightarrow y + 2x = -1$

Similar calculations lead to: $T(2,1)$, $m_{RT} = 1$

Equation of $RT$: $y - x = -1$

Solve: $\begin{cases} y + 2x = -1 \\ y - x = -1 \end{cases}$

subtract: $3x = 0 \Rightarrow x = 0$

Put $x = 0$ in $y - x = -1 \Rightarrow y = -1$

So $G(0, -1)$

**TOP TIP**

Always work with axes diagrams to make sure your answers make sense e.g. that gradients look right and points of intersection are where they are supposed to be!

## Quick Test 30

A triangle has vertices $A(-1, 2)$, $B(3, 4)$ and $C(3, -2)$.

**1.** Find the equation of the perpendicular bisector of

a) side $AB$ b) side $AC$.

**2.** Find the point of intersection $G$ of these two perpendicular bisectors.

**3.** Show that in triangle $ABC$ all three perpendicular bisectors are concurrent.

# The circle equation

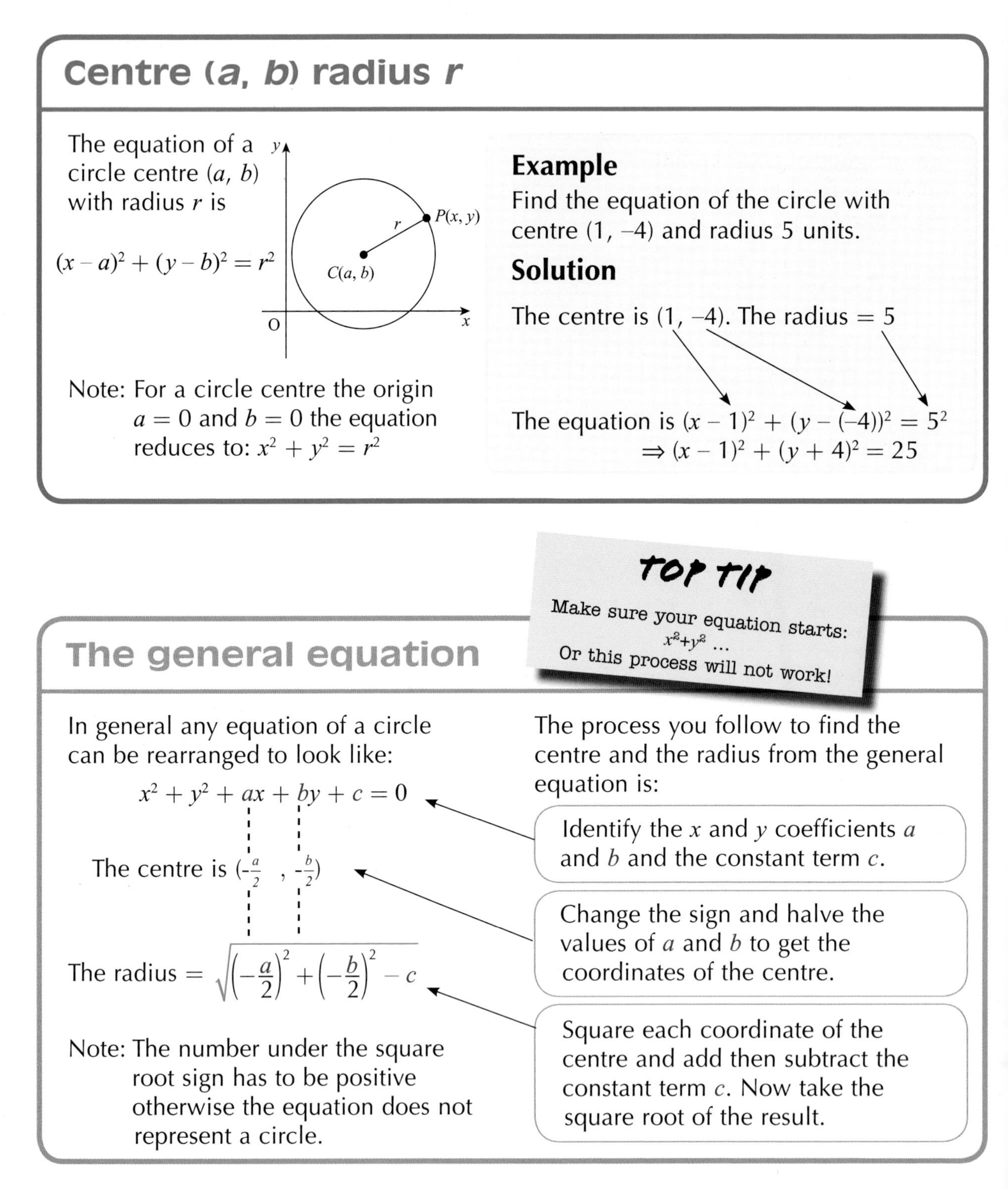

## Centre (*a*, *b*) radius *r*

The equation of a circle centre $(a, b)$ with radius $r$ is

$$(x - a)^2 + (y - b)^2 = r^2$$



Note: For a circle centre the origin $a = 0$ and $b = 0$ the equation reduces to: $x^2 + y^2 = r^2$

**Example**

Find the equation of the circle with centre (1, –4) and radius 5 units.

**Solution**

The centre is (1, –4). The radius = 5

The equation is $(x - 1)^2 + (y - (-4))^2 = 5^2$

$\Rightarrow (x - 1)^2 + (y + 4)^2 = 25$

**TOP TIP**

Make sure your equation starts: $x^2+y^2$ …
Or this process will not work!

## The general equation

In general any equation of a circle can be rearranged to look like:

$$x^2 + y^2 + ax + by + c = 0$$

The centre is $\left(-\frac{a}{2}, -\frac{b}{2}\right)$

The radius $= \sqrt{\left(-\frac{a}{2}\right)^2 + \left(-\frac{b}{2}\right)^2 - c}$

Note: The number under the square root sign has to be positive otherwise the equation does not represent a circle.

The process you follow to find the centre and the radius from the general equation is:

- Identify the $x$ and $y$ coefficients $a$ and $b$ and the constant term $c$.
- Change the sign and halve the values of $a$ and $b$ to get the coordinates of the centre.
- Square each coordinate of the centre and add then subtract the constant term $c$. Now take the square root of the result.

### Example

For what range of values of $k$ does $x^2 + y^2 - 2x + 6y + k = 0$ represent a circle?

### Solution

Centre is $(1, -3)$

Radius $= \sqrt{1^2 + (-3)^2 - k} = \sqrt{10 - k}$

So $10 - k > 0$, i.e. $10 > k$ or $\mathbf{k < 10}$ (or there's no circle!)

### Example

Find the centre and radius of the circle $x^2 + y^2 - 2x + 3y - 3 = 0$

### Solution

$$x^2 + y^2 - 2x + 3y - 3 = 0$$

Centre is $\left(1, -\frac{3}{2}\right)$

Radius $= \sqrt{1^2 + (-\frac{3}{2})^2 - (-3)}$

$\sqrt{1 + \frac{9}{4} + 3} = \sqrt{\frac{25}{4}} = \frac{5}{2}$

## Completing squares

It is possible to 'complete squares' to find the radius and centre from the general equation:

$$x^2 + y^2 - 4x + 5y + 1 = 0$$

$$x^2 - 4x + y^2 + 5y = -1$$

$$(x - 2)^2 + (y + \tfrac{5}{2})^2 = -1 + 4 + \tfrac{25}{4} = \tfrac{37}{4}$$

$x^2 - 4x + 4$ $\quad$ $y^2 + 5y + \frac{25}{4}$

The centre is $(2, -\frac{5}{2})$ and the radius $= \sqrt{\frac{37}{4}} = \frac{\sqrt{37}}{2}$

Rearrange to get $x$ terms and $y$ terms together.

Complete the squares. You will have introduced extra terms (in purple) on the left and need to add them on the right to balance the equation.

Now compare with:

$$(x - a)^2 + (y - b)^2 = r^2$$

Centre: $(a, b)$ $\quad$ Radius $= r$

# Quick Test 31

1. Find the equation of the circle with centre $(-3, 7)$ and radius $\sqrt{3}$
2. Find the centre and the radius of each circle:
   a) $(x - \frac{1}{2})^2 + (y + \frac{5}{2})^2 = 5$ $\quad$ b) $x^2 + y^2 - x - y - \frac{1}{2} = 0$
3. A circle with radius 2 units touches the $x$-axis and touches the $y$-axis. Give the four possible equations for the circle.
4. Show that the circle $x^2 + y^2 - 4x + 6y + 9 = 0$ touches the $y$-axis but does not intersect the $x$-axis.

# Lines and circles

## How many points of intersection?

To find points of intersection of a line and a circle you solve their equations simultaneously. This will result in a quadratic equation. Three distinct situations can arise...

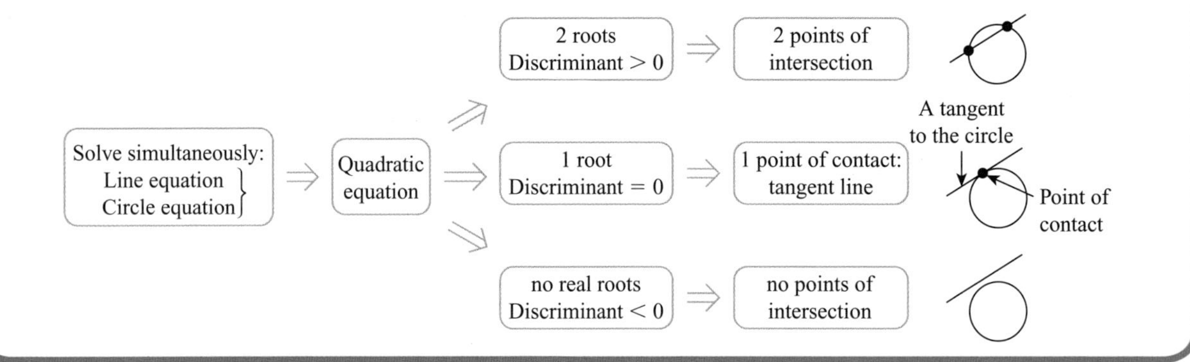


## Showing a line is a tangent

You need to show that when you solve the line equation and the circle equation simultaneously you only get one solution. You then can use this one value of $x$ to find the point of contact.

**Example**

Show that $y = 2x - 10$ is a tangent to the circle $x^2 + y^2 - 4x + 2y = 0$ and find the point of contact.

**Solution**

For the points of intersection solve:

$$\left.\begin{aligned} y &= 2x - 10 \\ x^2 + y^2 - 4x + 2y &= 0 \end{aligned}\right\}$$

Substitute $y = 2x - 10$ in the circle equation.

This gives... $x^2 + (2x - 10)^2 - 4x + 2(2x - 10) = 0$

$x^2 + 4x^2 - 40x + 100 - 4x + 4x - 20 = 0$

$5x^2 - 40x + 80 = 0 \Rightarrow 5(x^2 - 8x + 16) = 0 \Rightarrow 5(x - 4)(x - 4) = 0$

Since there is only one solution, $x = 4$, the line $y = 2x - 10$ is a tangent to the circle. Now let $x = 4$ in $y = 2x - 10 \Rightarrow y = 2 \times 4 - 10 = -2$. The point of contact is $(4, -2)$.

## Tangent at a given point

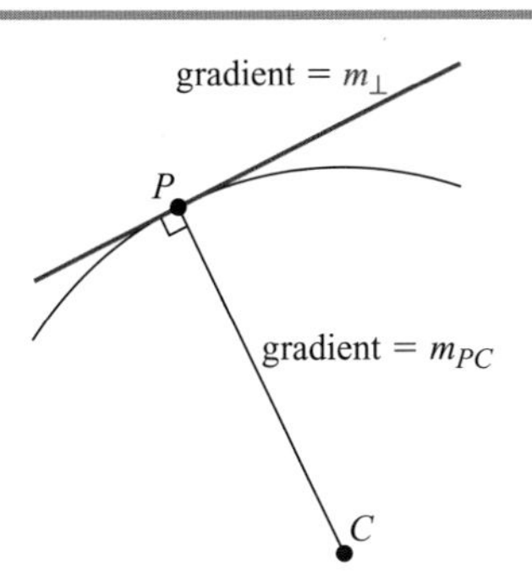


The tangent at $P$ is perpendicular to the radius $CP$ from the centre $C$ to the point of contact $P$.

So $m_\perp \times m_{pc} = -1$

**Example**

Find the equation of the tangent at the point $P(-4, 4)$ on the circle $(x + 2)^2 + y^2 = 20$.

**Solution**

The centre is $C(-2, 0)$ so $m_{cp} = \frac{4-0}{-4-(-2)} = \frac{4}{-2} = -2$

$\Rightarrow m_\perp = \frac{1}{2}$. This is the gradient of the tangent.

A point on the tangent is $(-4, 4)$.

The equation is $y - 4 = \frac{1}{2}(x - (-4))$

$\Rightarrow 2y - 8 = x + 4 \Rightarrow 2y - x = 12$

## Using the discriminant to find a tangent

A tangent has **one** point of intersection with a circle. By imposing a 'one root only' condition in some problems you can find the equation of an unknown tangent. This usually involves setting the discriminant of a quadratic equation to zero.

**Example**

$y = mx$ is a tangent to the circle
$(x + 2)^2 + y^2 = 3$
Find the possible values of $m$

**Solution**

The points of intersection can be found by solving:

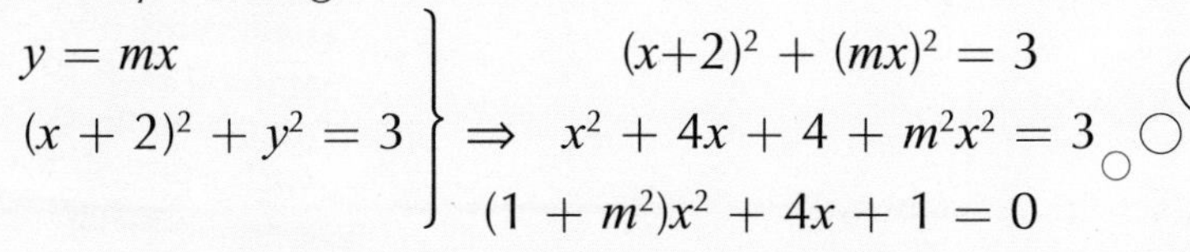

$$\left.\begin{array}{l} y = mx \\ (x+2)^2 + y^2 = 3 \end{array}\right\} \Rightarrow \begin{array}{l} (x+2)^2 + (mx)^2 = 3 \\ x^2 + 4x + 4 + m^2x^2 = 3 \\ (1 + m^2)x^2 + 4x + 1 = 0 \end{array}$$

This is a quadratic equation in $x$...
Discriminant $= 4^2 - 4(1 + m^2) \times 1$
('$b^2 - 4ac$') $= 16 - 4 - 4m^2$
$= 12 - 4m^2$

For the line to be a tangent the quadratic equation must have 1 root (or equal roots) so...
Discriminant = 0 giving $12 - 4m^2 = 0$

$\Rightarrow m^2 = 3 \Rightarrow m = \sqrt{3}$ or $-\sqrt{3}$

Check with a diagram. The circle has centre $(-2, 0)$ and radius $= \sqrt{3}$
The two possible tangents are shown.

y
x
$y = \sqrt{3}x$
$y = -\sqrt{3}x$

## Diametrically opposite points

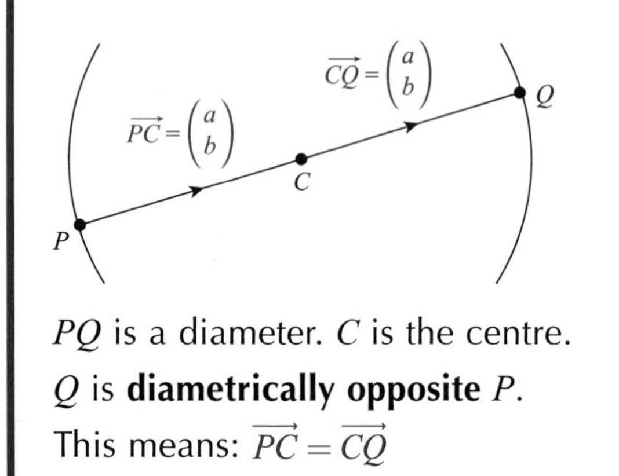


$PQ$ is a diameter. $C$ is the centre.

$Q$ is **diametrically opposite** $P$.

This means: $\overrightarrow{PC} = \overrightarrow{CQ}$

If you know two of the points then you can find the third point:

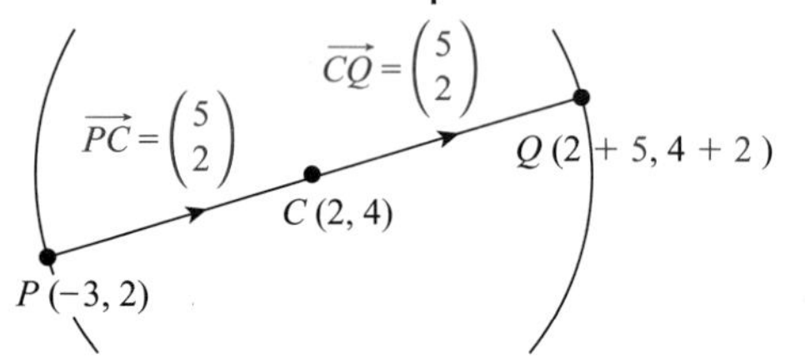


From $C$ to $Q$ you move 5 right and 2 up the same as from $P$ to $C$. So $Q$ is the point (7, 6).

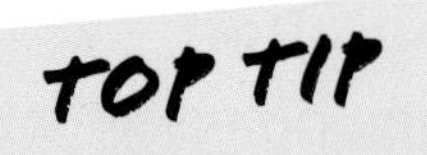


It is sensible to sketch the circles and lines before trying to solve the problem. It helps with your strategy!

## Quick Test 32

1. Show that the line $y - x = 5$ is a tangent to the circle $x^2 + y^2 + 6x + 4y + 5 = 0$ and find the coordinates of the point of contact.
2. Find the equation of the tangent to the circle $x^2 + y^2 - 16x + 35 = 0$ at the point (3, –4).
3. Find the possible values of $k$ if the line $y = x - k$ is a tangent to the circle $(x - 1)^2 + y^2 = 2$
4. $A(-1, 0)$ lies on circle $C_1$ with equation $(x - 2)^2 + (y - 1)^2 = 10$. $B$ is diametrically opposite $A$. Find the equation of circle $C_2$ with centre $B$ and with the same radius as circle $C_1$.

# Problem solving with circles

## Are the circles touching?

$D$ : distance between the two centres

$R$ : radius of large circle

$r$ : radius of small circle

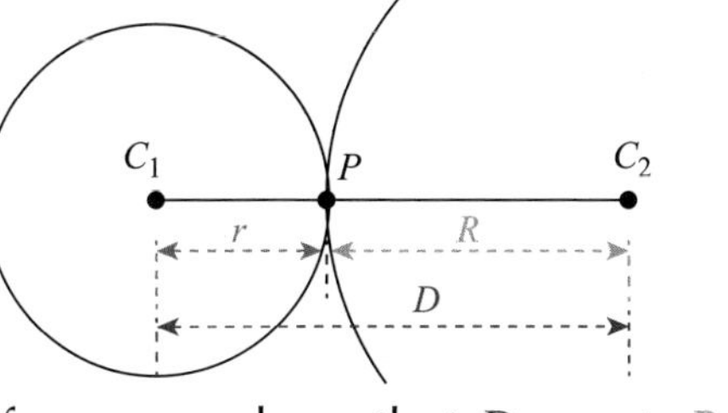


If you can show that $D = r + R$ then you have shown the two circles touch externally

If you can show that $D = R - r$ then you have shown the two circles touch internally

Notes: • If neither of these two equations is true then the circles do not touch (they could intersect in two points or miss each other).

• The point of contact $P$ can be found as the point which divides $C_1C_2$ in the ratio $r : R$ in the case of externally touching circles.

**Example**

Two circles have equations $(x - 3)^2 + (y - 1)^2 = 5$ and $(x + 3)^2 + (y + 2)^2 = 20$. Show that the circles touch and find the coordinates of the point of contact.

**Solution**

Circle 1: Centre is $C_1$ (3, 1) Radius $r = \sqrt{5}$

Circle 2: Centre is $C_2$ (–3, –2) Radius $R = \sqrt{20} = \sqrt{4 \times 5} = \sqrt{4} \times \sqrt{5} = 2\sqrt{5}$

$C_1C_2 = \sqrt{(-3-3)^2 + (-2-1)^2} = \sqrt{36+9} = \sqrt{45} = \sqrt{9 \times 5} = \sqrt{9} \times \sqrt{5} = 3\sqrt{5}$

$r + R = \sqrt{5} + 2\sqrt{5} = 3\sqrt{5} = C_1C_2$

So the two circles touch.

The point of contact $P$ divides $C_1C_2$ in the ratio 1:2

$\overrightarrow{PC_2} = 2\overrightarrow{C_1P} \Rightarrow \boldsymbol{c}_2 - \boldsymbol{p} = 2(\boldsymbol{p} - \boldsymbol{c}_1)$

$\Rightarrow 2\boldsymbol{p} + \boldsymbol{p} = 2\boldsymbol{c}_1 + \boldsymbol{c}_2 \Rightarrow 3\boldsymbol{p} = 2\boldsymbol{c}_1 + \boldsymbol{c}_2$

So $3\boldsymbol{p} = 2\begin{pmatrix}3\\1\end{pmatrix} + \begin{pmatrix}-3\\-2\end{pmatrix} = \begin{pmatrix}3\\0\end{pmatrix}$ giving $\boldsymbol{p} = \frac{1}{3}\begin{pmatrix}3\\0\end{pmatrix} = \begin{pmatrix}1\\0\end{pmatrix}$

Thus $P(1, 0)$

## What's the locus?

A **locus** (plural **loci**) is the path traced out by a moving point. Interesting loci are the ones that have restrictions imposed on the way the point moves.

**Example**

Find the equation of the locus of a point that moves so that it stays at a fixed distance of 3 units from the point (5, –3).

**Solution**

The point travels in a circle with radius 3 units and centre (5, –3) so its equation is $(x - 5)^2 + (y + 3)^2 = 9$

## Where's the centre?

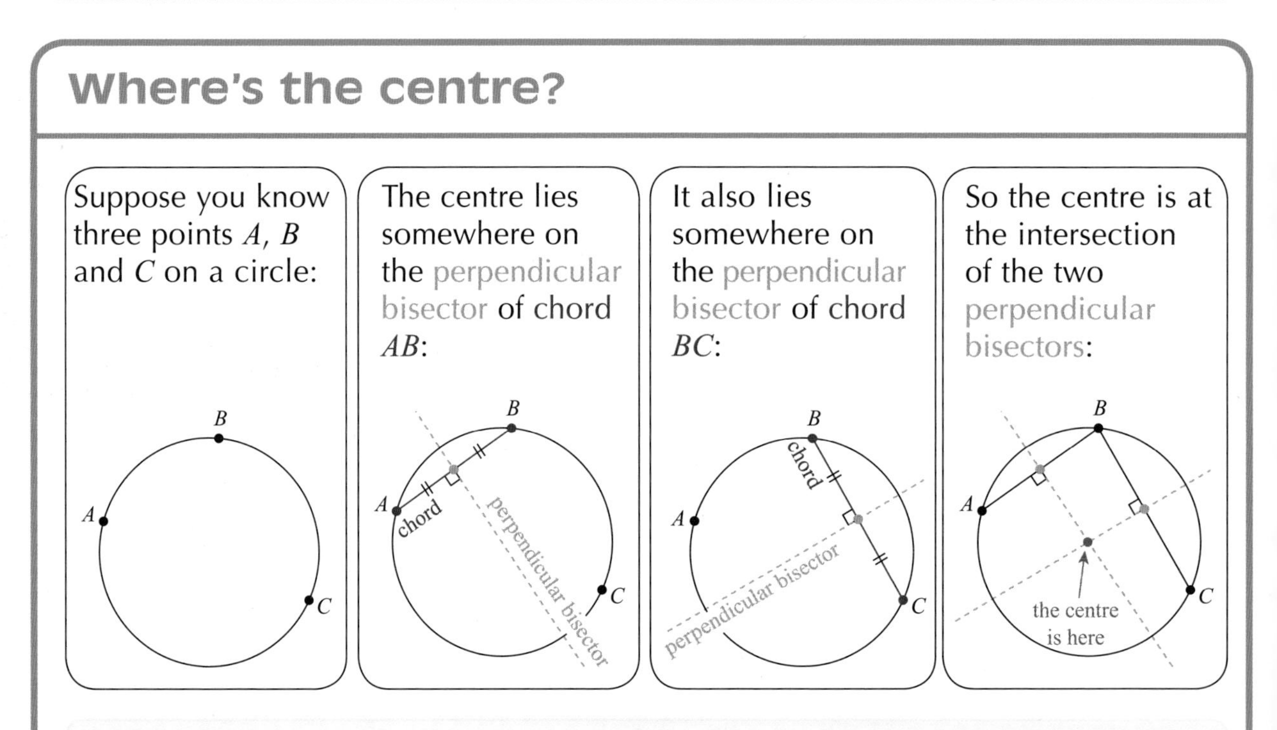


**Example**

Points $P(-3, 4)$, $Q(5, 4)$ and $R(9, -2)$ lie on a circle with centre $C$ and radius $r$ units.

a) Find the equation of the perpendicular bisector of chord $QR$.

b) Find the coordinates of the centre $C$.

c) Write down the value of $r$ and hence find the equation of the circle.

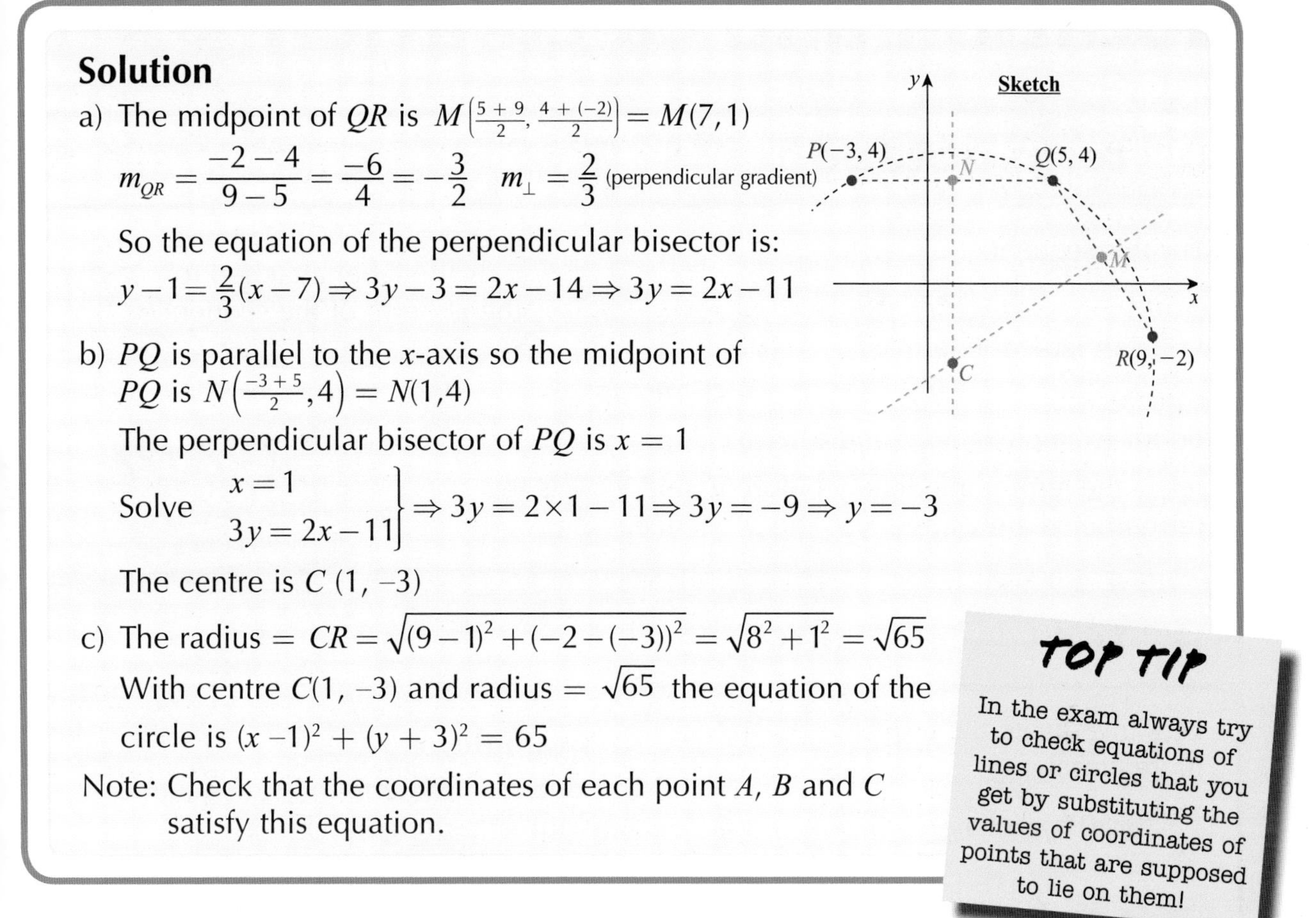

**Solution**

a) The midpoint of $QR$ is $M\left(\frac{5+9}{2}, \frac{4+(-2)}{2}\right) = M(7,1)$

$m_{QR} = \dfrac{-2-4}{9-5} = \dfrac{-6}{4} = -\dfrac{3}{2}$ $\quad m_{\perp} = \dfrac{2}{3}$ (perpendicular gradient)

So the equation of the perpendicular bisector is:

$y-1 = \frac{2}{3}(x-7) \Rightarrow 3y-3 = 2x-14 \Rightarrow 3y = 2x-11$

b) $PQ$ is parallel to the $x$-axis so the midpoint of $PQ$ is $N\left(\frac{-3+5}{2}, 4\right) = N(1,4)$

The perpendicular bisector of $PQ$ is $x = 1$

Solve $\left.\begin{array}{r} x=1 \\ 3y=2x-11 \end{array}\right\} \Rightarrow 3y = 2\times 1 - 11 \Rightarrow 3y = -9 \Rightarrow y = -3$

The centre is $C\ (1, -3)$

c) The radius $= CR = \sqrt{(9-1)^2 + (-2-(-3))^2} = \sqrt{8^2+1^2} = \sqrt{65}$

With centre $C(1, -3)$ and radius $= \sqrt{65}$ the equation of the circle is $(x-1)^2 + (y+3)^2 = 65$

Note: Check that the coordinates of each point $A$, $B$ and $C$ satisfy this equation.



**TOP TIP**

In the exam always try to check equations of lines or circles that you get by substituting the values of coordinates of points that are supposed to lie on them!

# Quick Test 33

1. Two circles have equations $x^2 + y^2 = 2$ and $x^2 + y^2 + 8x - 8y + 14 = 0$. Show that the circles touch and find the coordinates of the point of contact.
2. Find the equation of the locus of a point which stays $\sqrt{3}$ units from the point $(-2, -1)$.
3. Find the centre of the circle that passes through the points $A(-1, -1)$, $B(2, -1)$ and $C(2, 3)$.

# Notation and calculation

## $n^{th}$ term notation

For this sequence:

1, 3, 7, 15, 31, ...

you can label the terms:

$u_1, \quad u_2, \quad u_3, \quad u_4, \quad u_5, \quad \ldots \quad u_{n-1}, \quad u_n, u_{n+1}, \ldots$

So for example $u_3 = 7$ and $u_5 = 31$

$u_{n-1}$ is the term immediately before $u_n$ and $u_{n+1}$ is the term immediately after $u_n$ in the sequence.

$u_n$ is the label attached to the $n^{th}$ term of the sequence

**TOP TIP**

Sometimes $u_0$ is used for the 1st term in a sequence.

## $n^{th}$ term formulae

Let's investigate the sequence 1, 3, 7, ... more closely:

$u_1, \quad u_2, \quad u_3, \quad u_4, \quad \ldots \quad u_n$

1, 3, 7, 15 ...

$2^1 - 1, \; 2^2 - 1, \; 2^3 - 1, \; 2^4 - 1 \; \ldots \; 2^n - 1$

So $u_n = 2^n - 1$

This is an example of a $n^{th}$ term formula.

**Example**

Find the $10^{th}$ term in the sequence with $n^{th}$ term $u_n = 3n^2 - 2$

**Solution**

Substitute $n = 10$ in the formula:

$u_{10} = 3 \times 10^2 - 2 = \mathbf{298}$

## Recurrence relations

**TOP TIP**

The same recurrence relation can give different sequences if you use different 1st terms.

Here's another way to get the sequence 1, 3, 7, ...

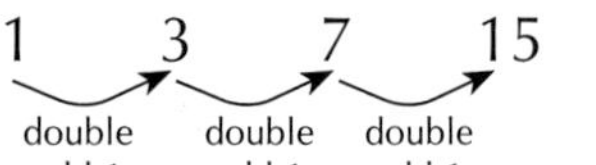


...

So $u_1 \to u_2$ (double add 1) gives $u_2 = 2u_1 + 1$

$u_2 \to u_3$ (double add 1) gives $u_3 = 2u_2 + 1$

**Example**

Find a recurrence relation and an $n^{th}$ term formula for the sequence 1, 3, 9, 27, ...

**Solution**

The 'build-up' rule is 'multiply by 3' so the recurrence relation is

$$\boldsymbol{u_{n+1} = 3u_n} \text{ with } u_1 = 1$$

(where $u_n$ is the $n^{th}$ term)

To find the $n^{th}$ term formula compare the terms with the powers of 3:

and in general

$u_n \to u_{n+1}$ (double, add 1) gives $u_{n+1} = 2u_n + 1$

This is a **recurrence relation**: it recurs as you build up the sequence.

| $u_1$ | $u_2$ | $u_3$ | $u_4$ |
|---|---|---|---|
| 1 | 3 | 9 | 27 |
| $3^0$ | $3^1$ | $3^2$ | $3^3$ |

one less each time

This gives the formula:

$$u_n = 3^{n-1}$$

## Linear recurrence relations

A **linear recurrence** relation is one with the form:

$$u_{n+1} = mu_n + c$$

$m$ is called the **multiplier**

$c$ is the constant term

A sequence generated by this type of recurrence relation is built up in this manner:

$u_1 \quad u_2 \quad u_3 \quad u_4 \quad \ldots$ (each step: $\times m$, $+c$)

So $u_2 = mu_1 + c$

$u_3 = mu_2 + c$

$u_4 = mu_3 + c$

and so on.

### Example

A sequence is defined by $u_{n+1} = au_n + b$ where $a$ and $b$ are constants. The first three terms are $u_1 = 20$, $u_2 = 15$ and $u_3 = 12{\cdot}5$.

Find the recurrence relation and hence calculate $u_5$

### Solution

20 → 15 ($u_1 \to u_2$, $\times a$, $+b$): $20a + b = 15$

15 → 12·5 ($u_2 \to u_3$, $\times a$, $+b$): $15a + b = 12{\cdot}5$

Solve: $20a + b = 15$, $15a + b = 12{\cdot}5$

Subtract: $5a = 2{\cdot}5$

$\Rightarrow a = 0{\cdot}5$

Put $a = 0{\cdot}5$ in $20a + b = 15$

so $20 \times 0{\cdot}5 + b = 15$

$\Rightarrow 10 + b = 15 \Rightarrow b = 5$

The recurrence relation is $u_{n+1} = 0.5u_n + 5$

So $u_4 = 0{\cdot}5u_3 + 5 = 0{\cdot}5 \times 12{\cdot}5 + 5 = 11{\cdot}25$

which gives $u_5 = 0{\cdot}5 \times 11{\cdot}25 + 5 = 10{\cdot}625$

# Quick Test 34

1. A recurrence relation is defined by $u_{n+1} = \frac{2}{3}u_n - 1$. If $u_1 = 27$, calculate $u_3$.
2. Evaluated $u_6$ if $u_n = 2^{5-n} + \frac{1}{2}$
3. The first three terms of a sequence generated by $u_{n+1} = pu_n + q$ are 2, 5 and 14. Calculate the values of the constants $p$ and $q$.

# Limits and context problems

## Conditions for a limit to exist

Sequences generated by the recurrence relation $u_{n+1} = mu_n + c$ behave in remarkably different ways as the sequence continues. The different types of behaviour depend entirely on the value of the multiplier $m$. These graphs are typical and show the values of $u_n$ as $n$ increases:

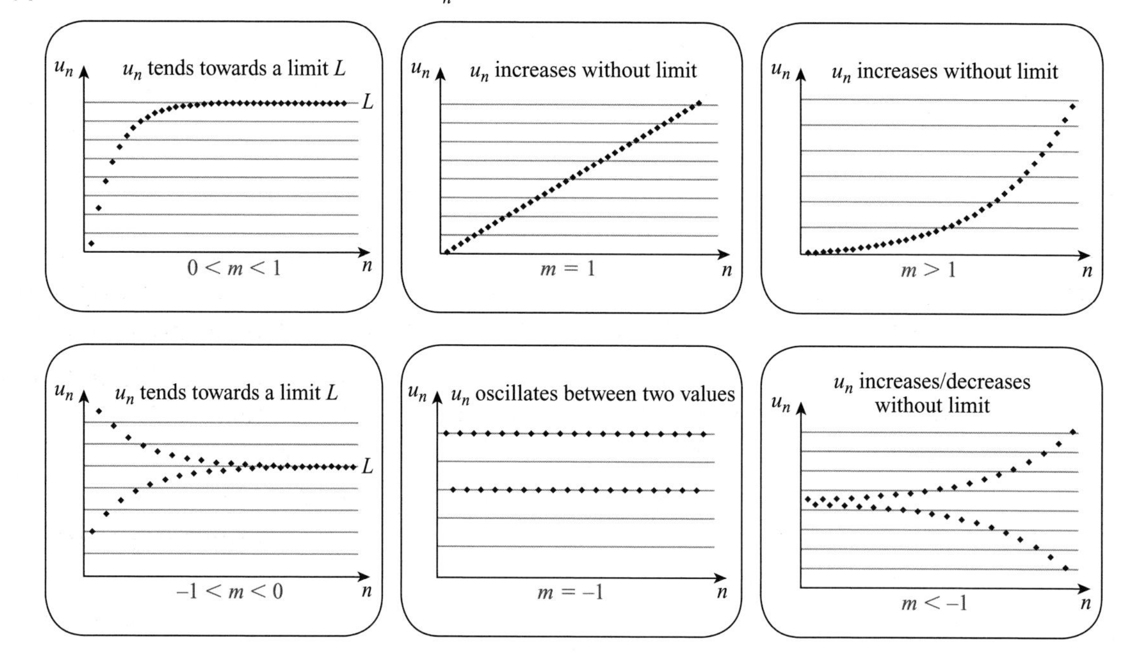


In two cases $0 < m < 1$ and $-1 < m < 0$ as $n$ gets larger and larger the values of $u_n$ get closer and closer to a limiting value $L$.

For a **limit to exist** the multiplier $m$ must lie between $-1$ and $1$ ($-1 < m < 1$).

## Finding the limit $L$

For the recurrence relation

$$u_{n+1} = mu_n + c$$

if $-1 < m < 1$ (the multiplier lies between $-1$ and $1$) then any sequence of values generated by this relation will eventually 'level out' at some limiting value $L$.

Putting this value $L$ into the recurrence relation will give $L$ again for the next term:

$$L = mL + c \quad \text{so} \quad L - mL = c \quad (1 - m)L = c$$

$$L = \frac{c}{1 - m} \quad (-1 < m < 1)$$

This is the **algebraic method** for calculating the limit.

**Example**

A sequence is defined by the recurrence relation $u_{n+1} = 0{\cdot}7u_n + 14$ with 1st term $u_1$

Explain why this sequence has a limit as $n$ tends to infinity. Find the **exact** value of this limit.

**Solution**

The multiplier 0·7 lies between –1 and 1 and so **a limit exists.**

Let the limit be $L$ then $L = 0{\cdot}7L + 14$

so $0{\cdot}3L = 14$ so $L = \dfrac{14}{0{\cdot}3}$ giving $L = \dfrac{140}{3} = \mathbf{46\frac{2}{3}}$

**TOP TIP**

When finding a limit $L$ remember always to show first that $-1 < m < 1$ before you proceed.

## Problems in context

Context problems that are solved using recurrence relations have many features in common. The following series of steps follows the course of a typical problem:

**Step 1** To solve the problem you will probably need to do a recurring calculation to produce a sequence of values. Try to calculate the first few values (eg 10 tonnes of pollutant, 1 week later 12·5 tonnes etc).

**Step 2** If $u_n$ is the $n^{th}$ term in this sequence of values then state clearly what meaning $u_n$ has in the given context (eg $u_n$ is number of tonnes of pollutant after $n$ weeks).

**Step 3** Describe the recurring calculation using $u_{n+1}$ and $u_n$ (eg $u_{n+1} = 0{\cdot}8u_n + 2$).

**Step 4** Use this recurrence relation to solve the problem (eg calculate a limit etc).

Note: Be very careful when finding the multiplier:

A reduction of 70% from $u_n$ to $u_{n+1}$ involves a multiplier of 0·3 since 30% remains of $u_n$.

**Example**

Two different types of water-purifying machines are in use.

Type A removes 35% of pollutants each day and is used in a tank which receives 15 litres of new pollutant at the end of each day Type B daily removes 60% of pollutants but is operating in a tank where 25 litres of new pollutant are dumped after each day's Operation.

In the long run which tank contains less pollutant?

**Solution**

After $n$ days let the amount of pollutant in tank A be $A_n$ litres and in tank B be $B_n$ litres. The recurrence relations which model this situation are:

$A_{n+1} = 0{\cdot}65A_n + 15$ and
$B_{n+1} = 0{\cdot}4B_n + 25$

In both cases the multipliers, namely 0·65 and 0·4, lie between −1 and 1 and so a limit exists in each case. Let the limit for tank A be $L$ litres and for tank B be $M$ litres, then:

$L = 0{\cdot}65L + 15$
$\Rightarrow L - 0{\cdot}65L = 15$
$\Rightarrow 0{\cdot}35L = 15$
$\Rightarrow L = \dfrac{15}{0{\cdot}35}$
$\doteqdot 42{\cdot}9$

$M = 0{\cdot}4M + 25$
$\Rightarrow M - 0{\cdot}4M = 25$
$\Rightarrow 0{\cdot}6M = 25$
$\Rightarrow M = \dfrac{25}{0{\cdot}6}$
$\doteqdot 41{\cdot}7$

In the long run tank B will contain 41·7 litres (to 3 sig figs) of pollutant, approximately 1·2 litres less than the amount of pollutant in tank A.

# Quick Test 35

1. For the sequence defined by each recurrence relation explain whether or not there is a limit as $n$ tends to infinity. If there is a limit then find its **exact** value.

   a) $u_{n+1} = 0{\cdot}4u_n + 1$ b) $u_{n+1} = \dfrac{u_n}{4} + 5$ c) $7u_{n+1} = 2u_n + 8$

2. The body destroys 70% of a drug in a day so a daily 28 unit injuection is then given. The initial injection was 50 units.
   a) At no time should the body contain more than 50 units of the drug. Is this course of treatment safe?
   b) Is it safe to increase the daily injections to 36 units?

# Applications of differentiation

## Intervals

In context problems the variable ($x$ or $t$ etc) is often restricted to a limited set of values.

For example the cardboard box shown might have its depth, $x$ cm, restricted to no less than 40 cm and no more than 60 cm.

So the interval of acceptable values is $40 \leq x \leq 60$

Other types of intervals are, for example: $t > 10$ or $v \leq 2{\cdot}3$

**TOP TIP**

You should try to sketch the graph of the function to see if your answers make sense. For Paper 2 questions this is a good use of a graphing calculator.

## Max/min values on an interval

For a graph $y = f(x)$ if the values of $x$ are resticted to an interval $a \leq x \leq b$ then the maximum and minimum values of $f$ will be found at the stationary points or at the end points of the interval:

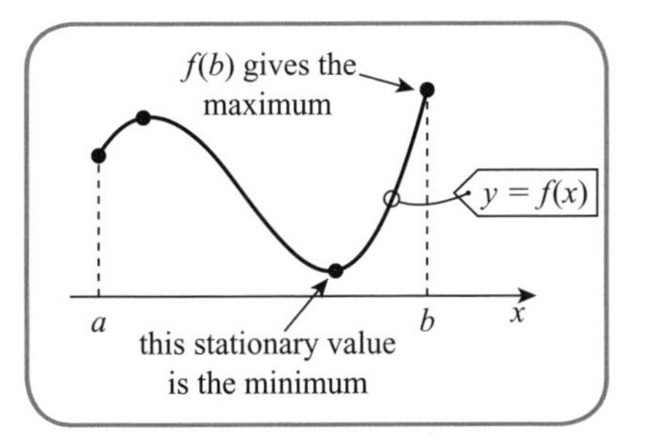


But look what happens for a different interval for the same graph:

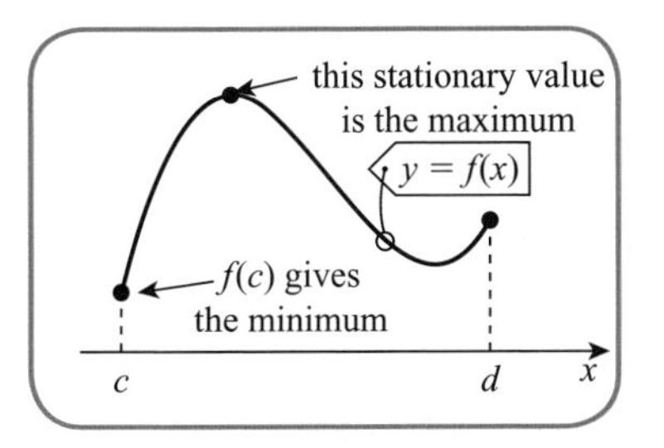


Note: The actual maximum or minimum value can be found by substituting the corresponding $x$-value into the formula $f(x)$.

**Example**

Find the maximum and minimum values of $f(x) = x^3 - 3x + 2$ on the interval $-2 \leq x \leq 3$

**Solution**

$f(x) = x^3 - 3x + 2$

$\Rightarrow f'(x) = 3x^2 - 3$

For stationary points set $f'(x) = 0$

So $3x^2 - 3 = 0$

$\Rightarrow 3(x^2 - 1) = 0$

$\Rightarrow 3(x - 1)(x + 1) = 0$

$\Rightarrow x - 1 = 0$ or $x + 1 = 0$

$\Rightarrow x = 1$ or $x = -1$

The stationary values are given by:

$f(1) = 1^3 - 3\times1 + 2 = 0$

$f(-1) = (-1)^3 - 3\times(-1) + 2 = -1 + 3 + 2 = 4$

The end points of the interval give values:

$f(-2) = (-2)^3 - 3 \times (-2) + 2 = -8 + 6 + 2 = 0$

$f(3) = 3^3 - 3 \times 3 + 2 = 27 - 9 + 2 = 20$

The maximum value is 20 (when $x = 3$)
The minimum value is 0 (when $x = 1$ or $-2$)

## Optimisation problems

The following example shows how finding stationary points can help to solve an **optimisation** problem:

### Example

The diagram shows a poster with width $x$ metres.

The total area of the poster is required to be $2\,m^2$.

The print area leaves margins 0·1 m at the top and bottom edges and 0·05 m margins at the left and right edges as shown.

Find the dimensions for the poster to maximise the print area.

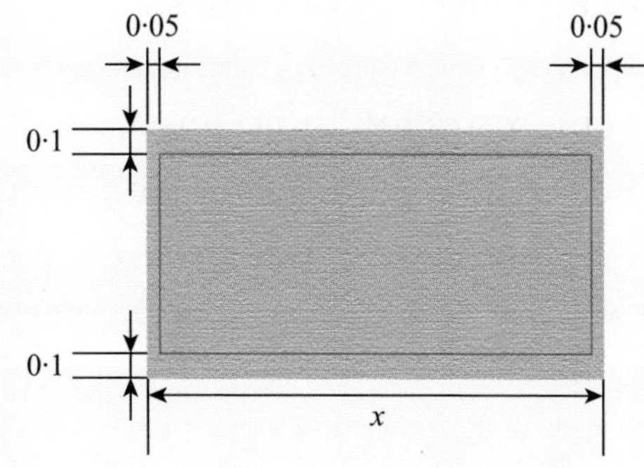


### Solution

width $\times\, x = 2 \Rightarrow$ width $= \frac{2}{x}$

so print area $A(x) = (x - 2\times0\cdot05)\left(\frac{2}{x} - 2\times0\cdot1\right) = (x - 0\cdot1)\left(\frac{2}{x} - 0\cdot2\right)$

$= 2 - 0\cdot2x - \frac{0\cdot2}{x} + 0\cdot02 = 2\cdot02 - 0\cdot2x - \frac{0\cdot2}{x} = 2\cdot02 - 0\cdot2x - 0\cdot2x^{-1}$

For stationary points set $A'(x) = 0$

This gives $A'(x) = -0\cdot2 + 0\cdot2x^{-2} = -0\cdot2 + \frac{0\cdot2}{x^2} = 0 \Rightarrow \frac{0\cdot2}{x^2} = 0\cdot2 \Rightarrow x^2 = 1$

Since $x > 0$ ($x$ measures a length) $x = 1$ is the only solution.

You now show this gives a maximum value:

| $x$: | | 1 | |
|---|---|---|---|
| $A'(x) = -0\cdot2 + \frac{0\cdot2}{x^2}$ | pos | 0 | neg |
| Shape of graph: | / | — | \ |

So $x = 1$ gives a maximum print area with:

length $= x = 1$ m and width $= \frac{2}{x} = \frac{2}{1} = 2$ m.

The required dimensions are 1 m × 2 m.

## Quick Test 36

1. Find the maximum and minimum values of $f(x) = x^4 - 2x^2 - 3$ on the interval $-0\cdot5 \leq x \leq 1\cdot5$

2. A cuboid is to be cut from a long wedge of wood as shown in the diagram. The wedge is in the shape of a prism with a right-triangular end with base 20 cm and height 10 cm. When cut, the cuboid has to have dimensions $x$ cm × $h$ cm × $6x$ cm.

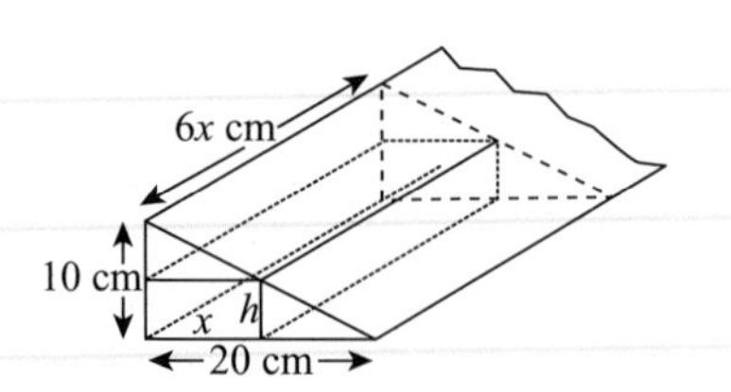


   a) Show that $h = 10 - \frac{1}{2}x$ (hint: use similar triangles).

   b) Show that the volume, $V\,cm^3$, of the cuboid is given by $V(x) = 60x^2 - 3x^3$

   c) Hence find the dimensions of the cuboid with the greatest volume that can be cut from the wedge.

# Applications of integration

## The area under a curve

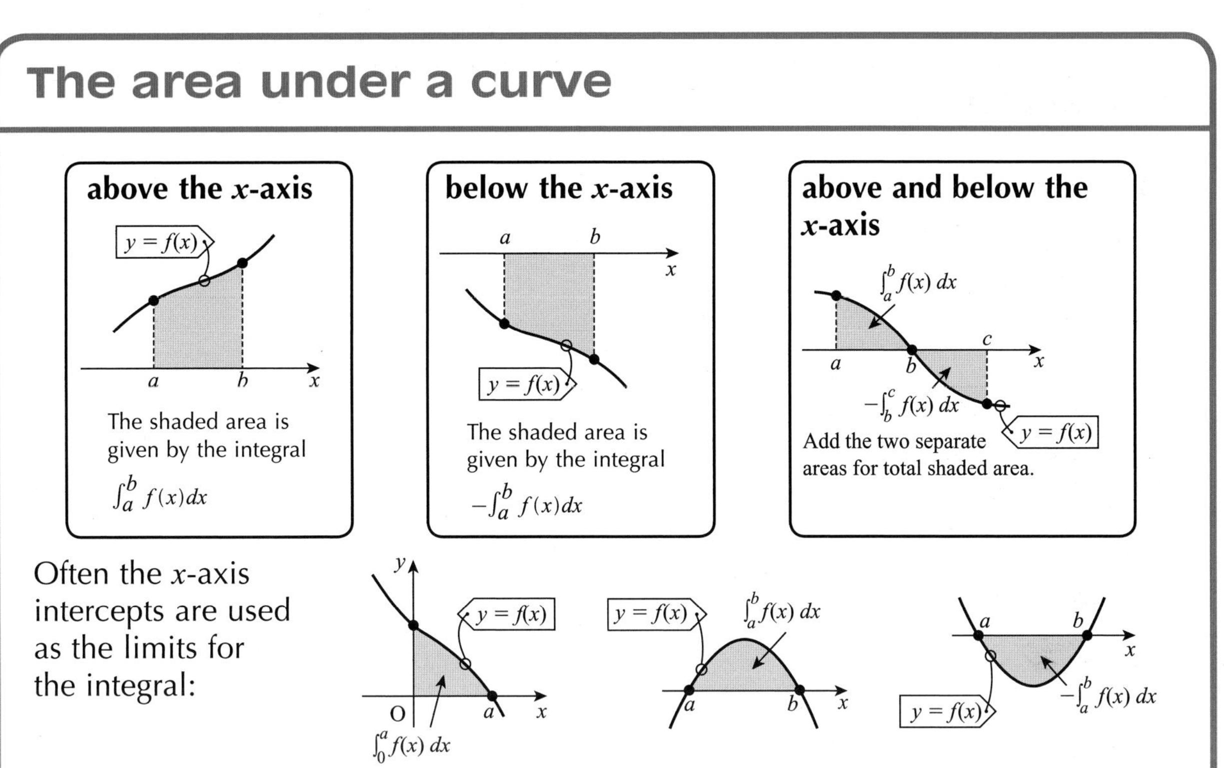


Often the $x$-axis intercepts are used as the limits for the integral:

### Example 1

Find the area enclosed by $y = x^2 + x - 2$ and the $x$-axis.

### Solution

For $x$-axis intercepts set $y = 0$

So solve $x^2 + x - 2 = 0$

$\Rightarrow (x + 2)(x - 1) = 0$

$\Rightarrow x = -2$ or $x = 1$

Now draw a sketch:

Calculating the integral gives:

$$\int_{-2}^{1}\left(x^2 + x - 2\right)dx = \left[\frac{x^3}{3} + \frac{x^2}{2} - 2x\right]_{-2}^{1} = \left(\frac{1^3}{3} + \frac{1^2}{2} - 2 \times 1\right) - \left(\frac{(-2)^3}{3} + \frac{(-2)^2}{2} - 2 \times (-2)\right)$$

$$= \frac{1}{3} + \frac{1}{2} - 2 - \left(-\frac{8}{3} + 2 + 4\right) = -\frac{9}{2} = -4\frac{1}{2}$$

Since the area is below the $x$-axis, to get the area change this value from negative to positive. So required area is **$4\frac{1}{2}$ unit²**.

### Note:

The unit of area used is a square of 1 unit along each axis.
You can sometimes estimate whether your answer makes sense!

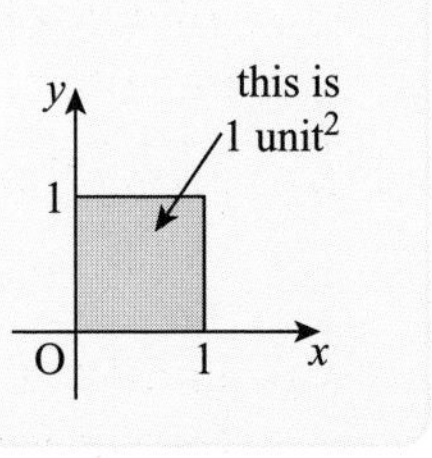

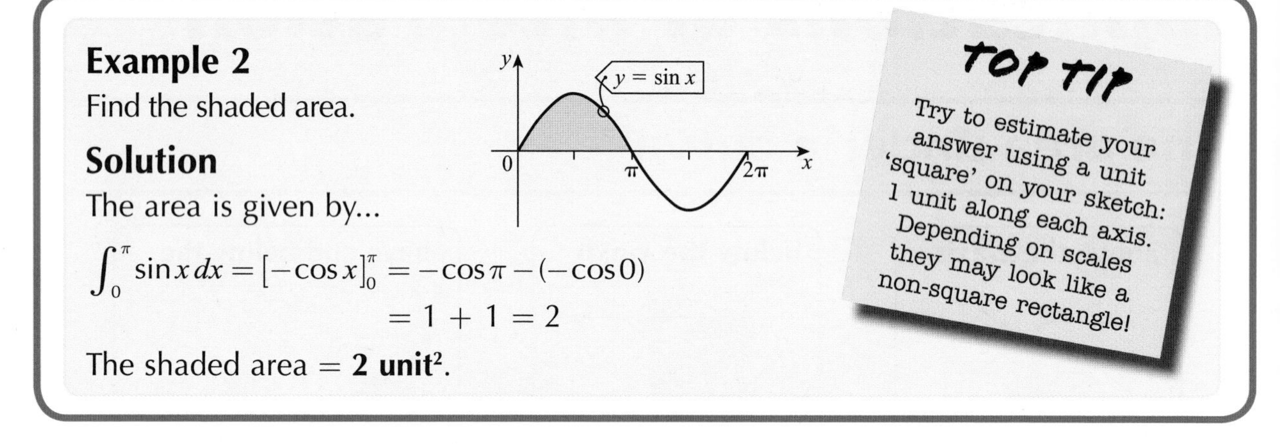


**Example 2**

Find the shaded area.

**Solution**

The area is given by...

$$\int_0^{\pi} \sin x\,dx = [-\cos x]_0^{\pi} = -\cos\pi - (-\cos 0)$$
$$= 1 + 1 = 2$$

The shaded area = **2 unit²**.

**TOP TIP**

Try to estimate your answer using a unit 'square' on your sketch: 1 unit along each axis. Depending on scales they may look like a non-square rectangle!

## The area between curves

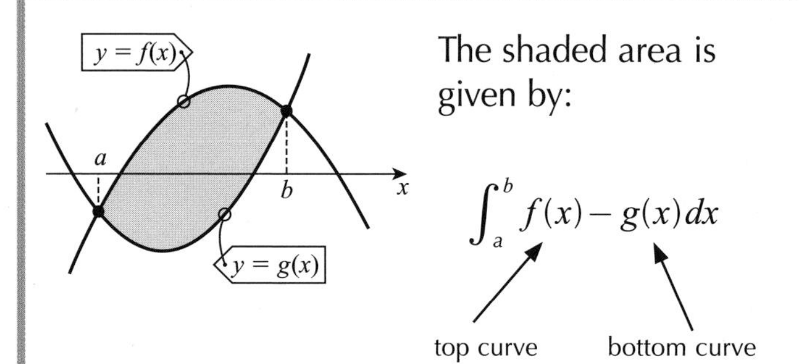


The shaded area is given by:

$$\int_a^b f(x) - g(x)\,dx$$

top curve     bottom curve

$x = a$ and $x = b$ are where the curves intersect.

Notes:

1. Simplify $f(x) - g(x)$ **first** before integrating... it's usually easier!
2. This result holds wherever the region is relative to the $x$-axis... above or below or even crossing the $x$-axis... use the same result.

**Example 1**

Find the area of the region enclosed by the line $y = x + 1$ and the parabola $y = 10 + 7x - 3x^2$

**Solution**

To find the points of intersection:

Solve:

$$\left.\begin{aligned} y &= x + 1 \\ y &= 10 + 7x - 3x^2 \end{aligned}\right\}$$

$$x + 1 = 10 + 7x - 3x^2$$
$$3x^2 - 6x - 9 = 0$$
$$3(x + 1)(x - 3) = 0$$
$$x = -1 \text{ or } x = 3$$

Make a sketch:

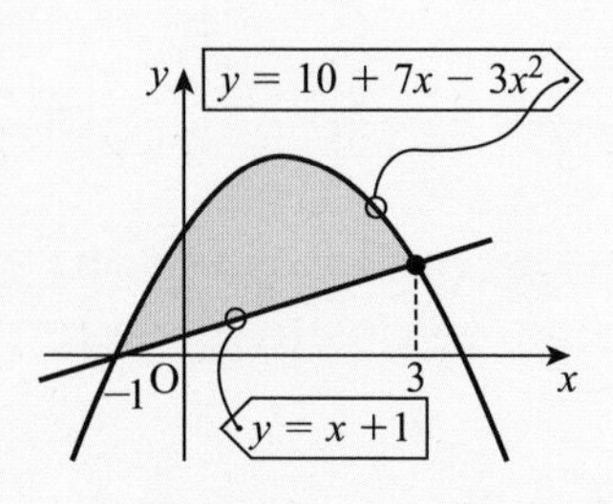


Shaded area

top graph     bottom graph

$$\int_{-1}^{3} (10 + 7x - 3x^2) - (x + 1)\,dx$$

$$\int_{-1}^{3} (10 + 7x - 3x^2 - x - 1)\,dx$$

$$\int_{-1}^{3}(9+6x-3x^2)\,dx = \left[9x+\frac{6x^2}{2}-\frac{3x^3}{3}\right]_{-2}^{1}$$

$$\left[9x+3x^2-x^3\right]_{-1}^{3}$$

$= (9\times3 + 3\times3^2 - 3^3) - (9 \times (-1) + 3 \times (-1)^2 - (-1)^3)$

$= 27 + 27 - 27 + 9 - 3 - 1 =$ **32 unit²**

## Example 2

Calculate the area enclosed by the curves $y = \sin x$ and $y = \sin 2x$ in the range $0 \leq x \leq \frac{\pi}{2}$ (shaded area in the diagram).

## Solution

For intersection points: solve: $\left.\begin{aligned} y &= \sin 2x \\ y &= \sin x \end{aligned}\right\}$

$\sin 2x = \sin x \Rightarrow 2\sin x\cos x - \sin x = 0$

$\sin x(2\cos x - 1) = 0 \Rightarrow \sin x = 0$ or $\cos x = \frac{1}{2}$

So $x = 0, \pi, 2\pi\ldots$ or $x = \frac{\pi}{2}, \frac{5\pi}{3}, \ldots$

The intersection you use is $x = \frac{\pi}{3}$:

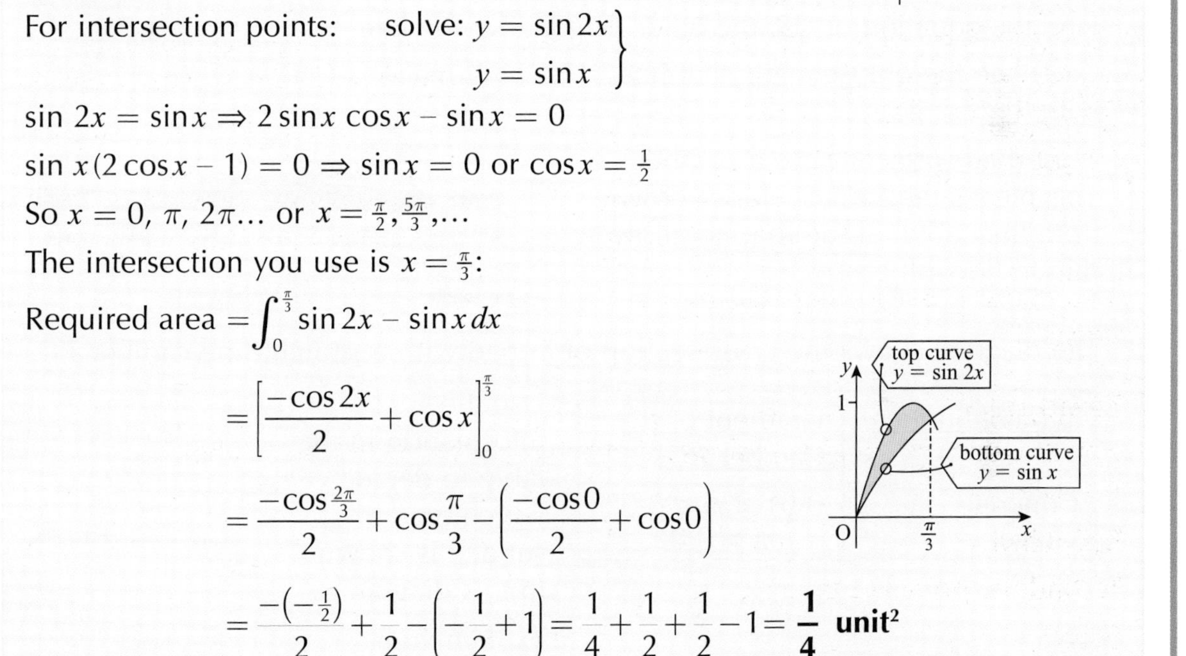


Required area $= \int_0^{\frac{\pi}{3}} \sin 2x - \sin x\,dx$

$$= \left[\frac{-\cos 2x}{2}+\cos x\right]_0^{\frac{\pi}{3}}$$

$$= \frac{-\cos\frac{2\pi}{3}}{2}+\cos\frac{\pi}{3}-\left(\frac{-\cos 0}{2}+\cos 0\right)$$

$$= \frac{-(-\frac{1}{2})}{2}+\frac{1}{2}-\left(-\frac{1}{2}+1\right)=\frac{1}{4}+\frac{1}{2}+\frac{1}{2}-1=\mathbf{\frac{1}{4}}\ \textbf{unit}^2$$

# Quick Test 37

**1.** Calculate the shaded area for each diagram:

a) b)

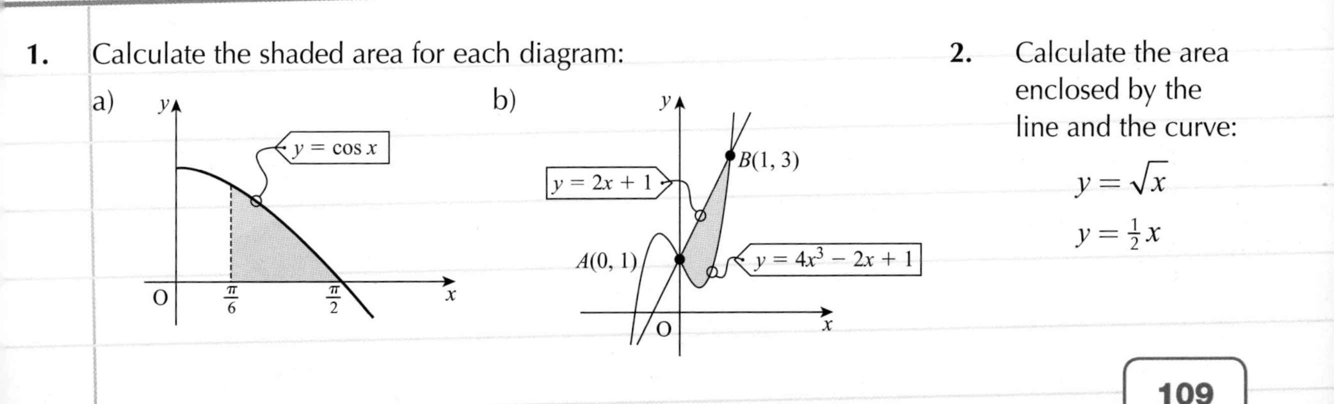


**2.** Calculate the area enclosed by the line and the curve:

$y = \sqrt{x}$

$y = \frac{1}{2}x$

# Sample Unit 3 test questions

## 1.1 Applying algebraic skills to rectilinear shapes, to circles and to sequences

1. Find the equation of the straight line parallel to $y - 5x = 2$ which passes through the point $(-1, 3)$.
2. $PQRS$ is a kite
   Diagonal $PR$ has equation $y = -\frac{1}{3}x + 2$
   Vertex $S$ is the point $(0,-3)$

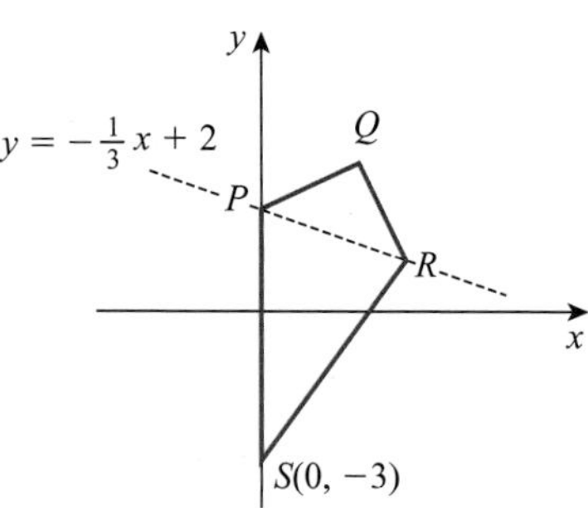


   Find the equation of $QS$, the other diagonal of the kite.
3. Calculate the obtuse angle that line $y = \frac{1}{2}x + 2$ makes with the $x$-axis.
4. Most road surfaces have a camber. This means the surface slopes down from the centre of the road to allow rain to flow off easily.

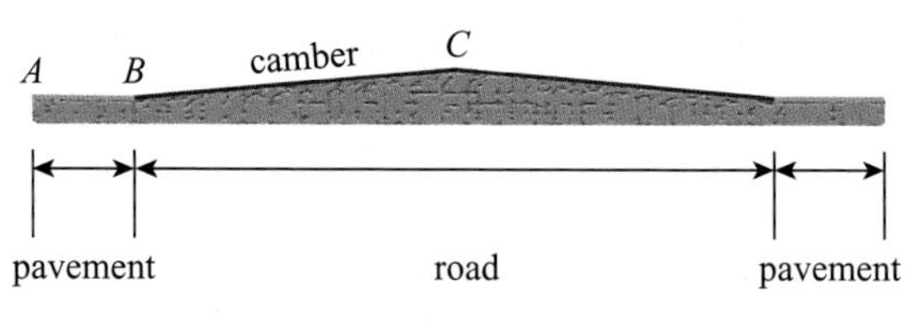


   The diagram shows the cross-section of a road with a straight-line camber $BC$.

| Type of Surface | Gradient of Camber $BC$ |
|---|---|
| Concrete | $0{\cdot}01 < m \leq 0.02$ |
| Gravel | $m > 0{\cdot}02$ |

   Use the information in the table to decide which type of surface is shown in this diagram:

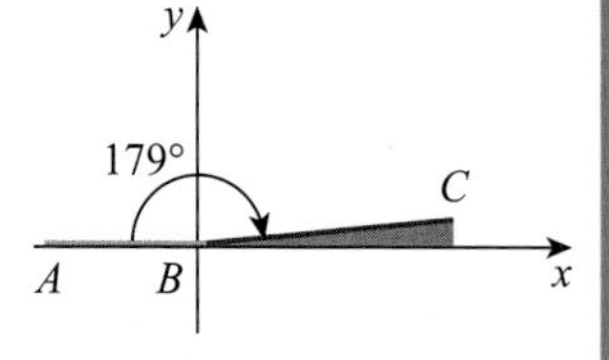


   Explain your answer fully.
5. Two congruent circles are shown. One touches the $x$-axis and the other touches the $y$-axis and passes through the point $(12, 0)$. Both circles have centres on the axes.
   Find the equation of the circle touching the $x$-axis. 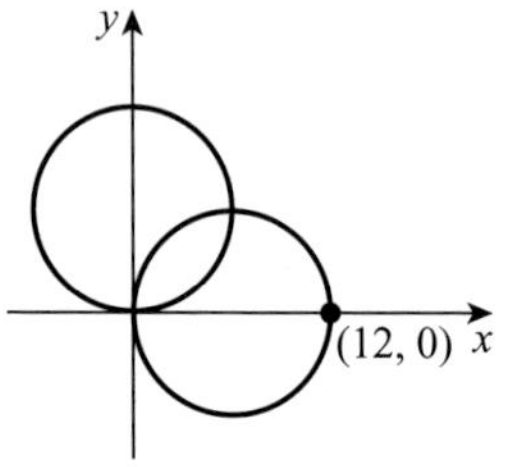
6. Ailsa says she can prove algebraically that $y = x - 1$ is a tangent to the circle with equation $(x - 1)^2 + (y + 4)^2 = 16$
   Can what she says be true? Show the reasoning for your answer.
7. The recurrence relation $u_{n+1} = au_n + b$ generates the sequence $u_1 = -2$, $u_2 = 1$ and $u_3 = 7$
   Find the values of $a$ and $b$ and hence calculate $u_5$.

8. Only 30% of a drug remains after a day in a patient's body.

   The patient is put on a course of daily 28 unit injections of the drug. The initial injection on the first day of the course was 50 units.

   Set up a recurrence relation showing the amount of drug in the patient's body immediately after an injection. Decide whether this course of injections is safe in the long run for the patient if the danger level is more than 50 units of the drug in the body.

## 1.2 Applying calculus skills to optimisation and area

1. The average manufacturing cost, £$C$, of a tablet screen protector depends on $x$, the number of thousands of the item that are produced per week.

   The formula is:

   $$C(x) = 3x^4 - 4x^3 + 5$$

   Find the value of $x$ which minimises the cost $C$.

2. Part of the graph $y = x(16 - x^2)$ is shown in the diagram. Calculate the shaded area.

3. The diagram shows the line $y = 5 - x$ and the curve $y = x^2 - 3x + 5$.

   The two graphs intersect where $x = 0$ and $x = 2$.

   Find the shaded area enclosed by the graphs.

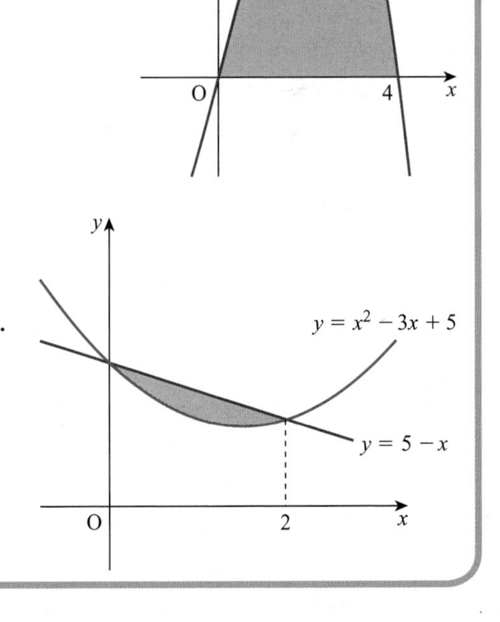

# Sample end-of-course exam questions (Unit 3)

## Non-calculator

1. $A(3, -1)$, $B(-1, 3)$ and $C(-2, 0)$ are the vertices of triangle $ABC$ as shown in the diagram. $M$ is the midpoint of $AB$. Find the equation of the line through $M$ perpendicular to $BC$.

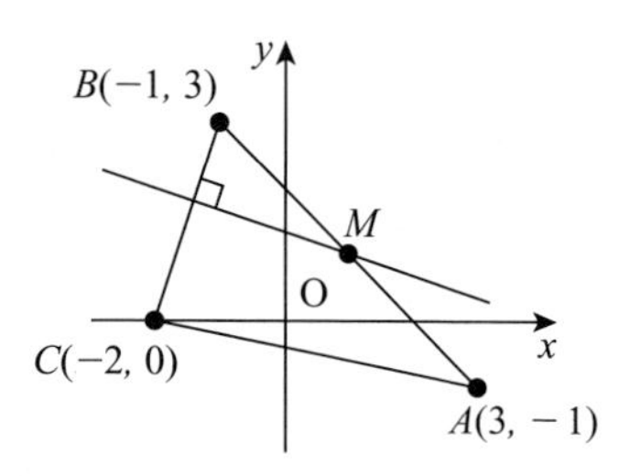


2. A sequence is defined by the recurrence relation $u_{n+1} = au_n + b$ with first term $u_1 = 5$.
   a) State the condition for this sequence to have a limit as $n$ tends to infinity.
   b) If $u_2 = 8$ and $u_3 = 9{\cdot}5$ calculate the values of $a$ and $b$.
   c) Find the exact value of the limit of this sequence as $n$ tends to infinity.

3. The diagram shows the design stage for a tambourine with four jingles. The line of centres of the tambourine (large circle) and the upper and lower jingles (small circles) is parallel to the $y$-axis. The centres of all four jingles lie on the circumference of the tambourine. The tambourine touches the $x$-axis and two of the jingles touch the $y$-axis as shown.

   If the equation of the tambourine (large circle) is $x^2 + y^2 - 2x - 6y + 1 = 0$ find the equation of the upper jingle (small circle).

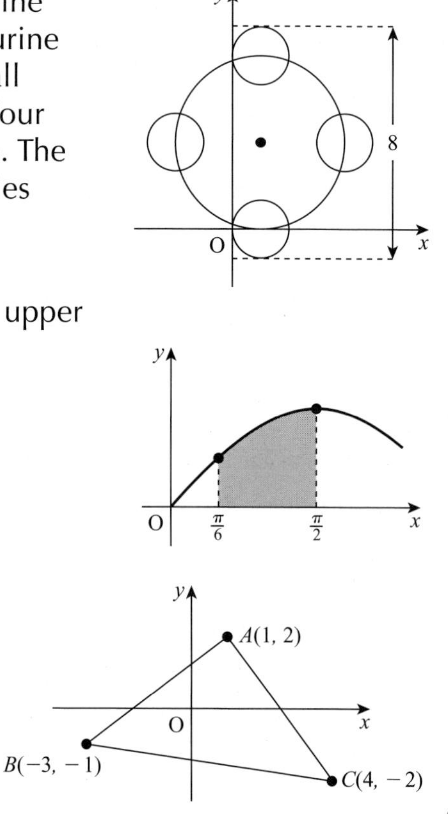


4. The diagram shows part of the graph $y = \sin x$
   Find the shaded area shown.

5. $A(1, 2)$, $B(-3, -1)$ and $C(4, -2)$ are the vertices of triangle $ABC$ as shown in the diagram.
   a) Show that triangle $ABC$ is isosceles.
   b) Side $AB$ makes an angle $\theta$ with the positive direction of the $x$-axis. Find the exact value of $\tan\theta$.

## Calculator allowed

1. $PQRS$ is a rhombus. $P$, $Q$ and $R$ have coordinates $P(-2, -1)$, $Q(-1, 4)$ and $R(4, 5)$. Find the equation of $SR$.

2. The diagram shows the cross-section of a wall of a hillside irrigation trench which has been modelled by the curve $y = x^3 - 2x + 1$.

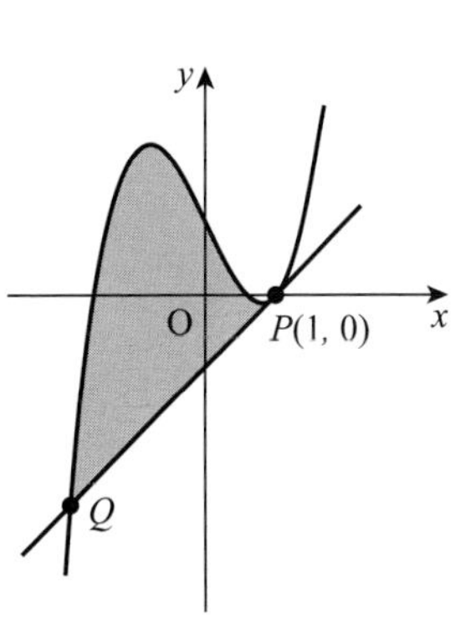


   The surface of the hillside has been modelled by $QP$, the tangent to the curve at $P(1, 0)$.

   a) Find the equation of the tangent $QP$.

   b) Find the coordinates of $Q$.

   c) Calculate the area of the shaded cross-section of the wall of the trench.

3. A metal component is in the form of a prism with cross-sectional area bounded by the curves with equations $y = 4\sqrt{x}$ and $y = 5 - x^2$ as shown in the diagram (all measurements are in centimetres). 30,000 of these components are to be produced.

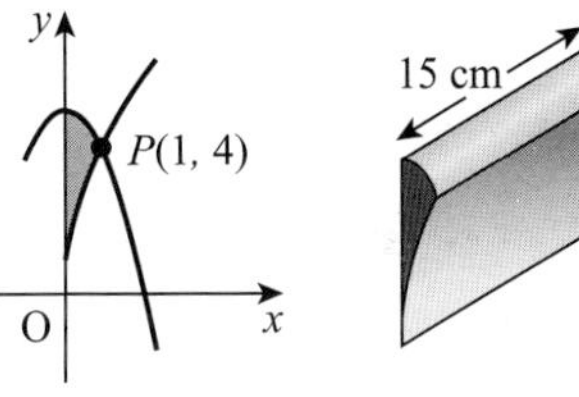


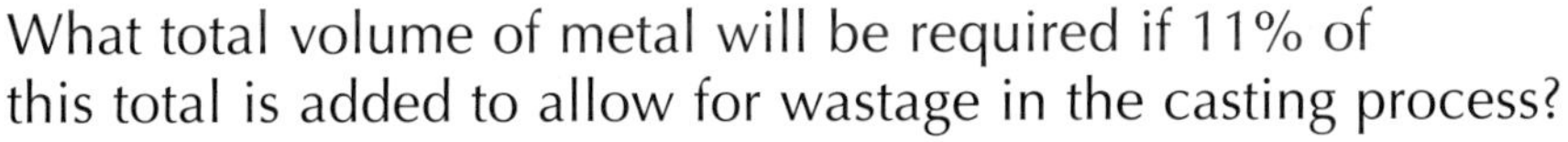

   What total volume of metal will be required if 11% of this total is added to allow for wastage in the casting process?

4. The diagram shows the circle with equation $x^2 + y^2 + 4x + 2y - 15 = 0$. A line with gradient 2 passes through $C$, the centre of the circle, and intersects the circle at points $A$ and $B$ as shown in the diagram.

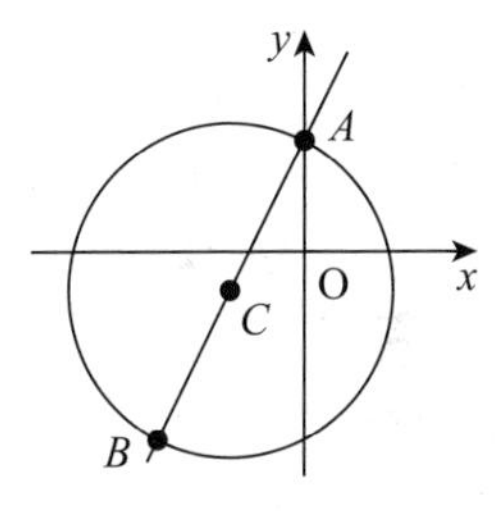


   a) Find the equation of the line $AB$ and hence show that $A$ lies on the $y$-axis.

   b) Find the equation of the tangent to the circle at the point $B$.

5. The diagram shows two identically sized circles that have line $AB$ as a common tangent with $T$ as the point of contact. The equation of $AB$ is $x = -2$.

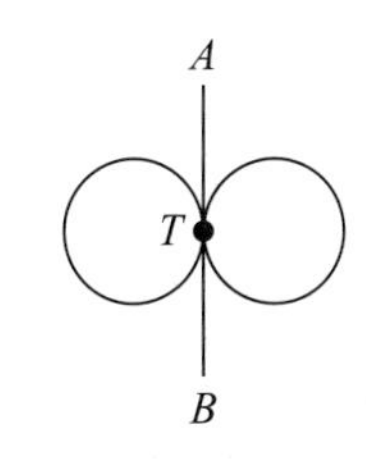


   a) One of the circles has equation $x^2 + y^2 - 2x + 2y - 7 = 0$.
   Find the equation of the other circle.

   b) A third circle is to be added to the figure in such a way that the other two circles lie inside it. The third circle has equation $x^2 + y^2 + 4x + 2y + c = 0$. Find the range of possible values of $c$.

# Quick Test Answers

## Quick Test 1

1. a) $x \in \mathbb{R}, x \neq -2$ b) $x \leq 10, x \in \mathbb{R}$
2. a) $b = 9$ b) 19

## Quick Test 2

1. a) $(3x - 1)^2$ b) $3x^2 - 1$ c) $9x - 4$ d) $x^4$
2. a) $f^{-1}(x) = 2x - 2$ b) $f^{-1}(x) = 7 - x$ c) $f^{-1}(x) = x^2 + 1$

## Quick Test 3

1. a) $2(x - 1)^2 + 4$, $a = 2$, $b = -1$, $c = 4$ b) $3(x - 1)^2 - 1$, $a = 3$, $b = -1$, $c = -1$
2. $k = \frac{1}{4}$, $m = 4$, $n = 2$

## Quick Test 4

1. 2.

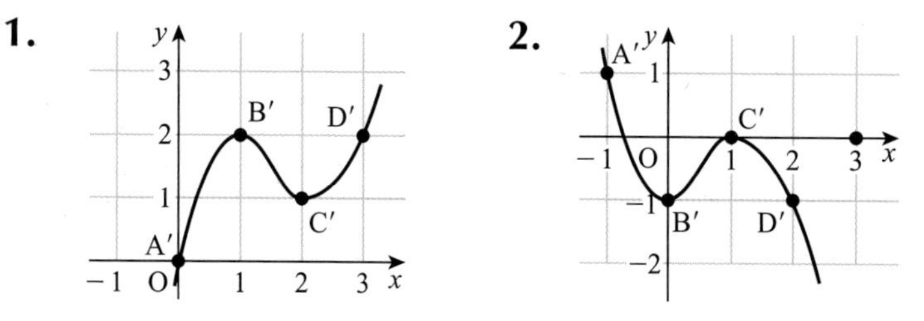


## Quick Test 5

1. a) $0 < a < 1$ b) $a = \frac{1}{3}$
2. £311,000 (to the nearest £1000)

## Quick Test 6

1. a) b) c) d)

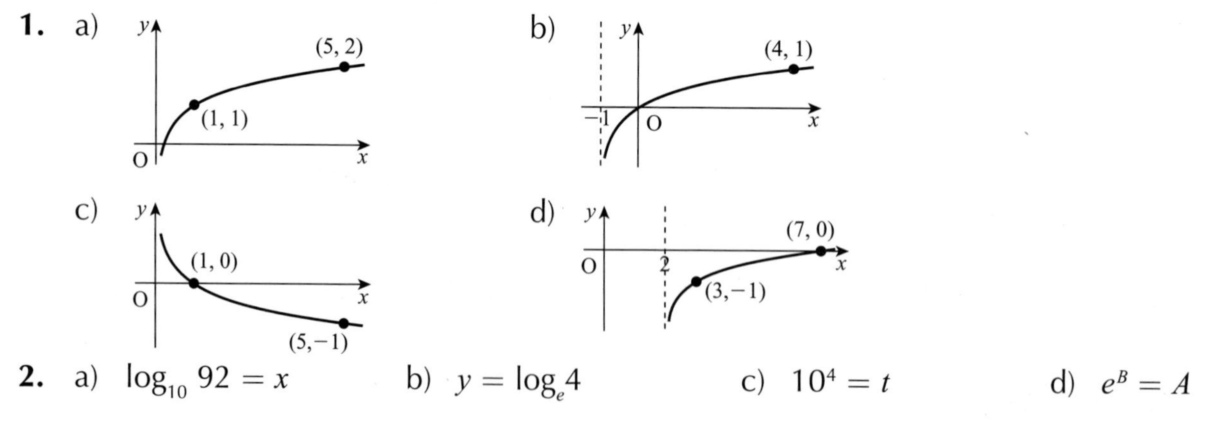


2. a) $\log_{10} 92 = x$ b) $y = \log_e 4$ c) $10^4 = t$ d) $e^B = A$

## Quick Test 7

1. 2
2. $\log_a x$
3. a) $x = 1{\cdot}16$ b) $x = 0{\cdot}83$
4. $a = \frac{1}{2}$ $b = 3$

## Quick Test 8

1. $\frac{3}{2}$
2. $0{\cdot}64$
3. $0,\ 2\pi$
4. a) $315°$ b) $\frac{\pi}{180} \times 40 \doteqdot 0{\cdot}698$

## Quick Test 9

1. $a = 4,\ b = 2$
2. a)

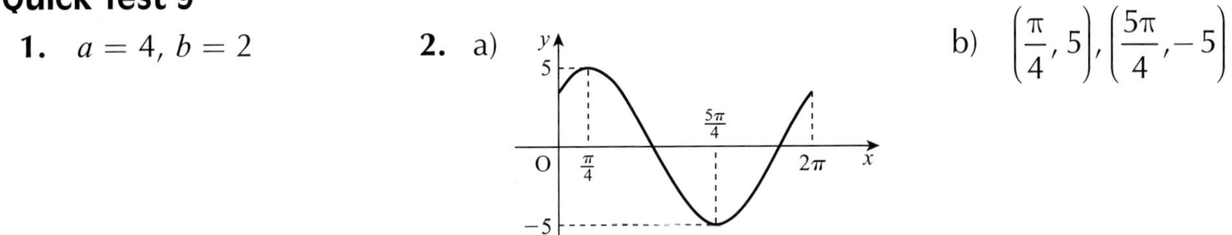


b) $\left(\frac{\pi}{4}, 5\right), \left(\frac{5\pi}{4}, -5\right)$

## Quick Test 10

1. $\frac{4}{\sin x°} = \frac{5}{\sin\frac{1}{2}(180 - x)°} \Rightarrow \frac{4}{\sin x°} = \frac{5}{\sin(90 - \frac{1}{2}x)°} \Rightarrow \frac{4}{\sin x°} = \frac{5}{\cos\frac{1}{2}x°} \Rightarrow \cos\frac{1}{2}x° = \frac{5}{4}\sin x°$

## Quick Test 11

1. $(\cos\theta + \sin\theta)(\cos\theta - \sin\theta) = \cos^2\theta - \sin^2\theta = \cos 2\theta$
2. $\frac{3}{5}$
3. $\cos(x + y) = \cos x \cos y - \sin x \sin y = \frac{8}{10} \times \frac{10}{5\sqrt{5}} - \frac{6}{10} \times \frac{5}{5\sqrt{5}} = \frac{8-3}{5\sqrt{5}} = \frac{5}{5\sqrt{5}} = \frac{1}{\sqrt{5}}$

## Quick Test 12

1. $\sqrt{10}\cos(x + 71{\cdot}6)°$
2. $2\sin(x + \frac{7\pi}{6})$

## Quick Test 13

1. a) $\begin{pmatrix} 2 \\ -4 \\ 3 \end{pmatrix}$ b) 9

## Quick Test 14

1. a) (i) $\begin{pmatrix} 4 \\ 5 \\ -4 \end{pmatrix}$ (ii) $\begin{pmatrix} 4 \\ 5 \\ -4 \end{pmatrix}$ b) parallelogram
2. $\overrightarrow{BC} = 3\overrightarrow{AB}$ so $AB \parallel BC$ with $B$ a shared point; $B$ divides $AC$ in the ratio $1:3$

## Quick Test 15

1. $33{\cdot}1°$ or $0{\cdot}577$
2. $m = -2$ or $3$
3. $\begin{pmatrix} -\frac{1}{3} \\ -\frac{2}{3} \\ \frac{2}{3} \end{pmatrix}$

## Quick Test 16

1. a) $|b| = |c| = \sqrt{2}$ b) 2
2. a) $M(0, 3, 2)$, $N(5, 2, 0)$ b) $5i - j - 2k$
3. a) $F_4 = \begin{pmatrix} -3 \\ -3 \\ 2 \end{pmatrix}$ b) $\sqrt{22}$

## Quick Test 17

1. $x^2 - 11x + 51 = 3x + 2 \Rightarrow x^2 - 14x + 49 = 0 \Rightarrow$ Discriminant $= (-14)^2 - 4 \times 1 \times 49 = 0$ $\Rightarrow$ one solution $\Rightarrow$ line is a tangent
2. $-\frac{3}{2} \leq k \leq \frac{3}{2}$
3. $-4 < x < 5$
4. $12x^2 + x - 1 = 0$
5. $\left(-\frac{1}{2}, -\frac{3}{4}\right), (3, 22)$

## Quick Test 18

1. $x^{-1}$
2. $f(3) = 0$ and $f(-2) = 0$
3. $4x^2 - 8x + 4$; rem $= -1$
4. $\frac{85}{81}$

## Quick Test 19

1. a) $f(1) = 0$ b) $f(x) = (x - 1)(2x^3 - x^2 - 2x + 1)$
   c) $2\times(-1)^3 - (-1)^2 - 2\times(-1) + 1 = 0$ d) $f(x) = (x - 1)^2 (x + 1)(2x - 1)$
2. $x = -\frac{1}{2}$, $x = \frac{2}{3}$ and $x = 1$

## Quick Test 20

1. a) 126·9, 306·9 b) 3·48, 5·94
2. a) $\frac{\pi}{6}, \frac{7\pi}{6}$ b) $\frac{2\pi}{3}, \frac{4\pi}{3}$
3. $\left(\frac{\pi}{6}, -\frac{1}{2}\right), \left(\frac{\pi}{2}, -\frac{1}{2}\right)$
4. a) 60, 300 b) $\frac{7\pi}{6}, \frac{3\pi}{2}, \frac{11\pi}{6}$
5. a) $g(x) = 2\sin(x - 30)°$ b) $x = 53{\cdot}6$ or $x = 186{\cdot}4$

## Quick Test 21

1. a) $3x^{-4}$ b) $-x^{-\frac{3}{2}}$ c) $\frac{1}{2}x^{-\frac{3}{2}}$
2. a) $\frac{3}{2}x^{-\frac{1}{2}} + 2x^{-2}$ b) $x^{-\frac{1}{2}} - \frac{1}{2}x^{-\frac{3}{2}}$
3. a) $\frac{1}{4}$ b) $\frac{13}{3}$
4. only $(-1, \frac{1}{2})$

## Quick Test 22

1. a) at $A$ and $C$ b) $A$: $3y + 3x = 5$, $B$: $y + 2x = 1$, $C$: $3y + 3x = 1$
2. $(-1, -1)$ maximum, $(0, -2)$ minimum

## Quick Test 23

1. 2.

## Quick Test 24

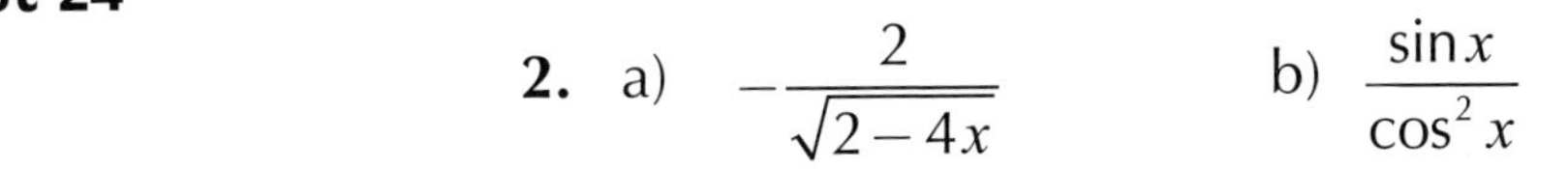

1. 2 2. a) $-\frac{2}{\sqrt{2-4x}}$ b) $\frac{\sin x}{\cos^2 x}$
3. $g'(x) = -\frac{1}{x^2}$ For $x \neq 0$: $\frac{1}{x^2} > 0$ so $-\frac{1}{x^2} < 0 \Rightarrow g'(x) < 0 \Rightarrow$ graph is decreasing

## Quick Test 25

1. $\frac{3\pi}{2}$
2. a) 400 m/min b) 0 m/min: max height reached
   c) –400 m/min: same as at start but in opposite direction (fallen back to ground)

## Quick Test 26

1. a) $-\frac{1}{2x^2} - \frac{2}{3}x^3 + C$ b) $\frac{5}{3}x^3 + 5x - \frac{3}{2x} + C$ c) $\frac{4}{3}x^{\frac{1}{2}} + \frac{2}{3}x^{\frac{3}{2}} + C$
2. a) $\frac{5}{4}x^4 + \frac{1}{2}x^2 - \cos x + C$ b) $2x^2 - \sin x + C$
3. $y = -\frac{1}{x} - \frac{x}{3} + 2$

## Quick Test 27

1. $\frac{20}{3}$ 2. $\frac{3\sqrt{2}-4}{4}$
3. a) $\sin^2 x = \frac{1}{2} - \frac{1}{2}\cos 2x$ b) $\frac{1}{2}x - \frac{1}{4}\sin 2x + C$ c) $\frac{\pi}{2}$

## Quick Test 28

1. $\frac{1}{3}$ 2. a) $\frac{3}{2}$; $(0, -2)$ b) $y = \frac{2}{3}x - 2$
3. $x = -2$

## Quick Test 29

1. a) $\frac{1}{2}$ b) $2y - x = -7$ c) $26{\cdot}6°$
2. $m_1 = -2$ and $m_2 = \frac{1}{2} \Rightarrow m_1 \times m_2 = -1 \Rightarrow$ lines are perpendicular. Intersection point is $(-1, 1)$

## Quick Test 30

1. a) $y + 2x = 5$ b) $y - x = -1$ 2. $G(2, 1)$
3. The 3rd equation is $y = 1$ and this passes through $G$ also

## Quick Test 31

1. $(x + 3)^2 + (y - 7)^2 = 3$ 2. a) $(\frac{1}{2}, -\frac{5}{2})$; $\sqrt{5}$ b) $(\frac{1}{2}, \frac{1}{2})$; 1
3. $(x - 2)^2 + (y - 2)^2 = 4$; $(x + 2)^2 + (y - 2)^2 = 4$; $(x + 2)^2 + (y + 2)^2 = 4$; $(x - 2)^2 + (y + 2)^2 = 4$
4. $y$-axis: set $x = 0 \Rightarrow y^2 + 6y + 9 = 0 \Rightarrow y = -3$ one solution so circle touches
   $x$-axis: set $y = 0 \Rightarrow x^2 - 4x + 9 = 0 \Rightarrow$ discriminant $< 0 \Rightarrow$ no solutions $\Rightarrow$ no intercepts

## Quick Test 32

1. Solving simultaneously gives one root $x = -5 \Rightarrow$ line is a tangent; contact point: $(-5, 0)$
2. $4y + 5x = -1$ 3. $k = -1$ or 3 4. $(x - 5)^2 + (y - 2)^2 = 10$

## Quick Test 33

1. Centres $(0, 0)$ and $(-4, 4)$ are $4\sqrt{2}$ units apart. The radii $\sqrt{2}$ and $3\sqrt{2}$ sum to $4\sqrt{2} \Rightarrow$ circles touch. Point of contact is $(-1, 1)$
2. $(x + 2)^2 + (y + 1)^2 = 3$ 3. $(\frac{1}{2}, 1)$

## Quick Test 34

1. $\frac{31}{3}$ 2. 1 3. $p = 3, q = -1$

## Quick Test 35

1. a) $m = 0{\cdot}4 \Rightarrow -1 < m < 1 \Rightarrow$ a limit $L$ exists: $L = \frac{5}{3}$

   b) $m = \frac{1}{4} \Rightarrow -1 < m < 1 \Rightarrow$ a limit $L$ exists: $L = \frac{20}{3}$

   c) $m = \frac{2}{7} \Rightarrow -1 < m < 1 \Rightarrow$ a limit $L$ exists: $L = \frac{8}{5}$

2. Let $u_n$ be amount of drug in body immediately after an injection $\Rightarrow u_{n+1} = 0{\cdot}3u_n + 28$ with $u_1 = 50$
   a) $m = 0{\cdot}3 \Rightarrow -1 < m < 1 \Rightarrow$ a limit $L$ exists: $L = 40 \Rightarrow$ course is safe
   b) $u_1 = 50$; $u_2 = 51$ so this is not safe

## Quick Test 36

1. max: $-2{\cdot}4375$ at $x = 1{\cdot}5$ min: $-4$ at $x = 1$

2. a) By similar triangles: $\frac{h}{10} = \frac{20 - x}{20} \Rightarrow h = 10 - \frac{1}{2}x$

   b) $V(x) = x \times \left(10 - \frac{1}{2}x\right) \times 6x = 60x^2 - 3x^3$

   c) Dimensions are: $\frac{40}{3}$ cm $\times$ $\frac{10}{3}$ cm $\times$ 80 cm

## Quick Test 37

1. a) $\frac{1}{2}$ unit$^2$ b) 1 unit$^2$ 2. $\frac{4}{3}$ unit$^2$

# Unit 1 sample test solutions

## 1.1 Applying algebraic skills to logarithims and exponentials

1. $\log_2 \dfrac{3ab}{3b} = \log_2 a$

2. $\log_b a^3 \times a^2 = \log_b a^5 = 5\log_b a$

3. Equivalent statement is: $2^1 = y - 3 \Rightarrow y = 5$

## 1.2 Applying trig skills to manipulating expressions

1. $\left.\begin{aligned} k\cos a &= 3 \\ k\sin a &= 1 \end{aligned}\right\} \Rightarrow \tan a = \frac{1}{3} \Rightarrow a \doteqdot 18{\cdot}4^\circ$ also $k = \sqrt{3^2 + 1^2} = \sqrt{10}$ giving $\sqrt{10}\cos(x - 18{\cdot}4)^\circ$

2. $1 - 2\sin x + \sin^2 x + 2\sin x = 1 + \sin^2 x = 1 + (1 - \cos^2 x) = 2 - \cos^2 x$

3. $\sin a \cos b + \cos a \sin b = \dfrac{9}{15} \times \dfrac{5}{13} + \dfrac{12}{15} \times \dfrac{12}{13} = \dfrac{189}{195} = \dfrac{63}{65}$

## 1.3 Applying algebra and trig skills to functions

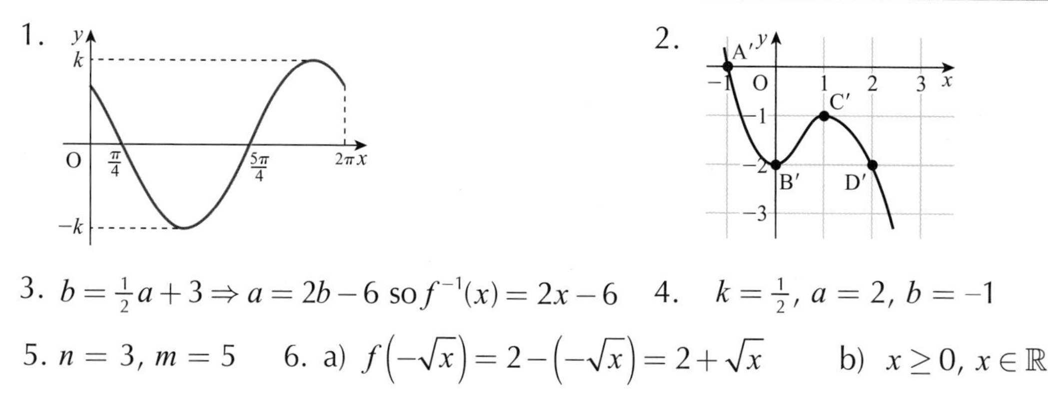


3. $b = \frac{1}{2}a + 3 \Rightarrow a = 2b - 6$ so $f^{-1}(x) = 2x - 6$

4. $k = \frac{1}{2},\ a = 2,\ b = -1$

5. $n = 3,\ m = 5$

6. a) $f\left(-\sqrt{x}\right) = 2 - \left(-\sqrt{x}\right) = 2 + \sqrt{x}$ b) $x \geq 0,\ x \in \mathbb{R}$

## 1.4 Applying geometric skills to vectors

1. a) $\overrightarrow{QR} = 3\overrightarrow{PQ}$, therefore collinear since $Q$ is a shared point

   b) Since $\overrightarrow{QR} = 3\overrightarrow{PQ}$ pipe 1 is three times as long as pipe 2

2. $\overrightarrow{AT} = \overrightarrow{2TB} \Rightarrow t - a = 2(b - t) \Rightarrow 3t = 2b + a = 2\begin{pmatrix} 7 \\ 3 \\ 1 \end{pmatrix} + \begin{pmatrix} 1 \\ -3 \\ 4 \end{pmatrix} = \begin{pmatrix} 15 \\ 3 \\ 6 \end{pmatrix} \Rightarrow t = \begin{pmatrix} 5 \\ 1 \\ 2 \end{pmatrix}$

   so $T(5, 1, 2)$

3. $\overrightarrow{DB} = \overrightarrow{DA} + \overrightarrow{AB} = -\begin{pmatrix} -4 \\ 2 \\ 3 \end{pmatrix} + \begin{pmatrix} -8 \\ 0 \\ 0 \end{pmatrix} = \begin{pmatrix} -4 \\ -2 \\ -3 \end{pmatrix}$

4. E, a, D, θ, b, F

   $a.b = 62,\ |a| = \sqrt{70},\ |b| = \sqrt{66},\ \dfrac{a.b}{|a| \times |b|} = 0{\cdot}912 \Rightarrow \theta \doteqdot 24{\cdot}2^\circ$

# Unit 2 sample test solutions

## 1.1 Applying algebraic skills to solve equations

1. Factors are $(x + 2)$, $(x + 3)$ and $(x - 1)$ so the roots are $x = -2$, $x = -3$ and $x = 1$

2. $2x^2 - x + a = 0$ has no solutions so discriminant $= (-1)^2 - 4 \times 2 \times a < 0$
   $\Rightarrow 1 - 8a < 0 \Rightarrow a > \frac{1}{8}$

3. $$\begin{array}{r|rrrr} -2 & 1 & -1 & -16 & -20 \\ & & -2 & 6 & 20 \\ \hline & 2 & -3 & -10 & 0 \end{array}$$
   so $f(x) = (x + 2)(x^2 - 3x - 10)$
   $= (x + 2)(x + 2)(x - 5)$
   $= (x + 2)^2(x - 5)$

## 1.2 Applying trig skills to solve equations

1. $\cos 3x = \frac{1}{2} \Rightarrow 3x = 60$ or $300$ or ... $\Rightarrow x = 20$ or $100$ or ...
   The only solutions in the range $0 < x \leq 120$ are $x = 20, 100$

2. $\cos\theta° - 5 \times 2\sin\theta°\cos\theta° = 0 \Rightarrow \cos\theta°(1 - 10\sin\theta°) = 0$
   $$\left\{\begin{array}{l} \Rightarrow \cos\theta° = 0 \text{ giving } \theta = 90 \\ \Rightarrow \sin\theta° = \frac{1}{10} \text{ giving } \theta \doteqdot 5{\cdot}7 \text{ (1st quadrant only)} \end{array}\right.$$

3. $\sin(\theta - 36{\cdot}9)° = \frac{2{\cdot}75}{5} = 0{\cdot}55 \Rightarrow \theta - 36{\cdot}9 \doteqdot 33{\cdot}4 \Rightarrow \theta \doteqdot 70{\cdot}3$ (1st quadrant only)

## 1.3 Applying calculus skills of differentiation

1. $f'(x) = -2 \times (-\sin x) = 2\sin x$

2. $y = \frac{5x^{\frac{1}{2}}}{x^1} - 2x^{\frac{1}{2}} = 5x^{-\frac{1}{2}} - 2x^{\frac{1}{2}} \Rightarrow \frac{dy}{dx} = -\frac{5}{2}x^{-\frac{3}{2}} - x^{-\frac{1}{2}}$

3. $\frac{dy}{dx} = x + 3$ so when $x = 2$: $\frac{dy}{dx} = 5$. Also when $x = 2$: $y = \frac{1}{2} \times 2^2 + 3 \times 2 - 2 = 6$
   So a point on the tangent is $(2, 6)$ and the gradient $= 5$:
   equation is $y - 6 = 5(x - 2) \Rightarrow y = 5x - 4$

4. a) $v = \frac{dh}{dt} = 16 - 8t$ so when $t = 0$ then $v = 16$ m/sec
   b) when $t = 2$ then $v = 16 - 8 \times 2 = 0$: it has reached its maximum height

## 1.4 Applying calculus skills of integration

1. $\frac{1}{2}\sin x + C$

2. $g(x) = \int (5 + x)^{-2}\, dx = \frac{(5 + x)^{-1}}{-1} + C = -\frac{1}{5 + x} + C$

3. $\int x^{-\frac{1}{2}} - 2x^{\frac{1}{2}}\, dx = \frac{x^{\frac{1}{2}}}{\frac{1}{2}} - \frac{2x^{\frac{3}{2}}}{\frac{3}{2}} + C = 2x^{\frac{1}{2}} - \frac{4}{3}x^{\frac{3}{2}} + C$

4. $\left[\frac{(x + 2)^4}{4}\right]_{-2}^{2} = \frac{(2 + 2)^4}{4} - \frac{(-2 + 2)^4}{4} = \frac{4^4}{4} - 0 = 4^3 = 64$

# Unit 3 sample test solutions

## 1.1 Applying algebraic skills to rectilinear shapes, to circles and to sequences

1. gradient = 5 so the equation is $y - 3 = 5(x + 1) \Rightarrow y = 5x + 8$
2. $m_{PR} = -\frac{1}{3} \Rightarrow m_{\perp} = 3$ so $m_{QS} = 3$ and a point on $QS$ is $(0, -3)$.
   The equation is $y + 3 = 3(x - 0) \Rightarrow y = 3x - 3$
3. The acute angle $= \tan^{-1}\frac{1}{2} \doteqdot 26{\cdot}6°$ so the obtuse angle $= 180° - 26{\cdot}6° = 153{\cdot}4°$
4. $BC$ makes a 1° angle with the $x$-axis $\Rightarrow m = \tan 1° = 0{\cdot}017... \Rightarrow 0{\cdot}01 < m \le 0{\cdot}02$
   $\Rightarrow$ it's concrete
5. radius = 6 and centre is (0, 6) $\Rightarrow$ equation is $(x - 0)^2 + (y - 6)^2 = 6^2 \Rightarrow x^2 + (y - 6)^2 = 36$
6. Solving the equations simultaneously gives:
   $(x - 1)^2 + ((x - 1) + 4)^2 = 16 \Rightarrow x^2 - 2x + 1 + x^2 + 6x + 9 = 16 \Rightarrow 2x^2 + 4x - 6 = 0$
   $\Rightarrow x^2 + 2x - 3 = 0 \Rightarrow (x + 3)(x - 1) = 0 \Rightarrow x = -3$ or $x = 1$
   Since there are two solutions the line cannot be a tangent so what she said was false.
7. $1 = a \times (-2) + b$ and $7 = a \times 1 + b$ so solve: $\left.\begin{array}{r} -2a+b=1 \\ a+b=7 \end{array}\right\} \Rightarrow a = 2,\ b = 5$
   So $u_{n+1} = 2u_n + 5$. The sequence is $-2, 1, 7, 19, 43, \ldots$ So $u_5 = 43$.
8. Let $u_n$ be the amount of the drug in the patient's body immediately after the $n^{th}$ injection.
   Then $u_{n+1} = 0{\cdot}3u_n + 28$ with $u_1 = 50$.
   The multiplier $m = 0{\cdot}3$ with $-1 < m < 1$ so a limit $L$ exists.
   So $L = 0{\cdot}3L + 28 \Rightarrow 0{\cdot}7L = 28 \Rightarrow L = 40$ which is safe being less than the danger level of 50 units.

## 1.2 Applying calculus skills to optimisation and area

1. $C'(x) = 12x^3 - 12x^2 = 12x^2(x - 1)$. For stationary values set $C'(x) = 0 \Rightarrow x = 0$ or $x = 1$
   Since $x > 0$ only consider $x = 1$. Nature table gives:

| $x$: | | 1 | |
|---|---|---|---|
| $C'(x) = 12x^2(x - 1)$: | neg | 0 | pos |
| Shape of graph: | \ | — | / |

   So $x = 1$ gives a minimum value for $C$
2. $\int_0^4 16x - x^3\,dx = \left[8x^2 - \frac{x^4}{4}\right]_0^4 = (8\times 4^2 - \frac{4^4}{4}) - (0 - 0) = 128 - 64 = 64$ unit$^2$
3. Shaded area
   $= \int_0^2 (5 - x) - (x^2 - 3x + 5)dx = \int_0^2 2x - x^2\,dx = \left[x^2 - \frac{x^3}{3}\right]_0^2 = 2^2 - \frac{2^3}{3} = \frac{4}{3}$ unit$^2$

# Unit 1 sample exam question solutions

## Non-calculator

1. $h(3\times\frac{\pi}{12}) = h(\frac{\pi}{4}) = \sin(2\times\frac{\pi}{4}) = \sin\frac{\pi}{2} = 1$

2. $x = \frac{\pi}{6}$ so $\sin x = \frac{1}{2}$ and $\cos x = \frac{\sqrt{3}}{2}$

$\Rightarrow \sin 2x = 2\sin x\cos x = 2\times\frac{1}{2}\times\frac{\sqrt{3}}{2} = \frac{\sqrt{3}}{2}$

3. a) Since AP : PC = 2 : 1

C(1, 4, 6)
P 1 bit
2 bit
A(4, −2, 0)

$$\overrightarrow{AP} = 2\overrightarrow{PC}$$
$$\Rightarrow \mathbf{p} - \mathbf{a} = 2(\mathbf{c} - \mathbf{p})$$
$$\Rightarrow \mathbf{p} - \mathbf{a} = 2\mathbf{c} - 2\mathbf{p}$$
$$\Rightarrow \mathbf{p} + 2\mathbf{p} = 2\mathbf{c} + \mathbf{a}$$
$$\Rightarrow 3\mathbf{p} = 2\mathbf{c} + \mathbf{a}$$

So

$$3\mathbf{p} = 2\begin{pmatrix}1\\4\\6\end{pmatrix} + \begin{pmatrix}4\\-2\\0\end{pmatrix} = \begin{pmatrix}6\\6\\12\end{pmatrix}$$

$$\Rightarrow \mathbf{p} = \frac{1}{3}\begin{pmatrix}6\\6\\12\end{pmatrix} = \begin{pmatrix}2\\2\\4\end{pmatrix}$$

So $P$ (2, 2, 4)

b) $\overrightarrow{BP} = \mathbf{p} - \mathbf{b} = \begin{pmatrix}2\\2\\4\end{pmatrix} - \begin{pmatrix}3\\5\\0\end{pmatrix} = \begin{pmatrix}-1\\-3\\4\end{pmatrix}$

So $\overrightarrow{BP} = -\mathbf{i} - 3\mathbf{j} + 4\mathbf{k}$

4. $2(\sqrt{3}\cos x - \sin x) = k\cos(x + a)$

$\Rightarrow 2\sqrt{3}\cos x - 2\sin x$

$= k\cos x\cos a - k\sin x\sin a$

So $k\cos a = 2\sqrt{3}$, $k\sin a = 2$ — Since cos $a$ and sin $a$ are positive, $a$ is in the 1st quadrant.

$$\frac{k\sin a}{k\cos a} = \frac{2}{2\sqrt{3}}$$

$$\Rightarrow \frac{\sin a}{\cos a} = \frac{1}{\sqrt{3}}$$

$$\Rightarrow \tan a = \frac{1}{\sqrt{3}}$$

$$\Rightarrow a = \frac{\pi}{6}$$

2, $\frac{\pi}{6}$, $\sqrt{3}$, $\frac{\pi}{3}$, 1

Also

$$(k\sin a)^2 + (k\cos a)^2 = 2^2 + (2\sqrt{3})^2$$
$$\Rightarrow k^2\sin^2 a + k^2\cos^2 a = 4 + 12$$
$$\Rightarrow k^2(\sin^2 a + \cos^2 a) = 16$$
$$\Rightarrow k^2\times 1 = 16 \Rightarrow k^2 = 16$$
$$\Rightarrow k = 4\ (k > 0)$$

So $2(\sqrt{3}\cos x - \sin x)$

$$= 4\cos\left(x + \frac{\pi}{6}\right)$$

5. a)

$$\overrightarrow{AB} = \mathbf{b} - \mathbf{a} = \begin{pmatrix}k\\k\\0\end{pmatrix} - \begin{pmatrix}1\\-2\\-k\end{pmatrix} = \begin{pmatrix}k-1\\k+2\\k\end{pmatrix}$$

$$\overrightarrow{AC} = \mathbf{c} - \mathbf{a} = \begin{pmatrix}4\\-3\\3-k\end{pmatrix} - \begin{pmatrix}1\\-2\\-k\end{pmatrix} = \begin{pmatrix}3\\-1\\3\end{pmatrix}$$

since $\overrightarrow{AB}$ and $\overrightarrow{AC}$ are perpendicular

then $\overrightarrow{AB}.\overrightarrow{AC} = 0 \Rightarrow \begin{pmatrix}k-1\\k+2\\k\end{pmatrix}\cdot\begin{pmatrix}3\\-1\\3\end{pmatrix} = 0$

$$\Rightarrow 3(k - 1) - 1(k + 2) + 3k = 0$$
$$\Rightarrow 3k - 3 - k - 2 + 3k = 0$$
$$\Rightarrow 5k - 5 = 0$$
$$\Rightarrow 5k = 5 \Rightarrow k = 1$$

b) $A(1, -2, -1)$, $C(4, -3, 2)$ and $D(13, -6, 11)$

$\overrightarrow{AC} = \begin{pmatrix} 3 \\ -1 \\ 3 \end{pmatrix}$ from part (*a*) above

$$\overrightarrow{CD} = \boldsymbol{d} - \boldsymbol{c} = \begin{pmatrix} 13 \\ -6 \\ 11 \end{pmatrix} - \begin{pmatrix} 4 \\ -3 \\ 2 \end{pmatrix}$$

$$= \begin{pmatrix} 9 \\ -3 \\ 9 \end{pmatrix}$$

So $\overrightarrow{CD} = 3\overrightarrow{AC}$ so $\overrightarrow{CD}$ and $\overrightarrow{AC}$ are parallel and since $C$ is a shared point then $A$, $C$ and $D$ are collinear

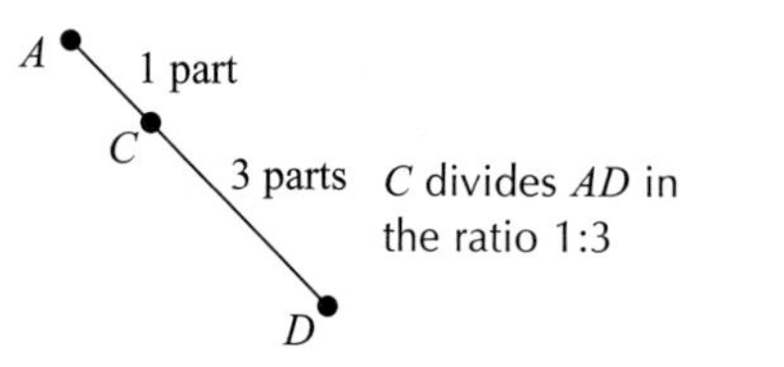


6. $a = 2$ (number of cycles in interval $0 \le x \le 2\pi$)
   $b = -1$ (graph $y = 2\sin 2x$ moved down 1 unit)

7. a) $\log_{\sqrt{a}} b = 2c$

$$\Rightarrow \left(\sqrt{a}\right)^{2c} = b$$

$$\Rightarrow (a^{\frac{1}{2}})^{2c} = b \Rightarrow a^{\frac{1}{2}\times 2c} = b$$

$$\Rightarrow a^c = b \Rightarrow \log_a b = c$$

b) By part (a) above, $\log_{\sqrt{a}} b$ has twice the value of $\log_a b$ ($2c$ is twice $c$).
   Since $\log_5 7$ is equal to $\log_{\sqrt{25}} 7$ it therefore has twice the value of $\log_{25} 7$ so:

$$\log_5 7 - \log_{25} 7 = 2\log_{25} 7 - \log_{25} 7$$

$$= \log_{25} 7 \text{ as required.}$$

## Calculator allowed

1. a) $\dfrac{2x^2 - 7x + 6}{x^2 - 4}$

$$= \frac{(2x-3)(x-2)}{(x-2)(x+2)} = \frac{2x-3}{x+2}$$

b) $\log_3(2x^2 - 7x + 6) - \log_3(x^2 - 4) = 2$

$$\Rightarrow \log_3\left(\frac{2x^2 - 7x + 6}{x^2 - 4}\right) = 2$$

$$\Rightarrow \log_3\left(\frac{2x-3}{x+2}\right) = 2$$

$$\Rightarrow \frac{2x-3}{x+2} = 3^2$$

$$\Rightarrow 2x - 3 = 9(x+2)$$

$$\Rightarrow 2x - 3 = 9x + 18$$

$$\Rightarrow -3 - 18 = 9x - 2x$$

$$\Rightarrow -21 = 7x$$

$$\Rightarrow x = \frac{-21}{7} = -3$$

2. a) $\overrightarrow{CA} = \boldsymbol{a} - \boldsymbol{c} = \begin{pmatrix} 3 \\ 2 \\ 5 \end{pmatrix} - \begin{pmatrix} 6 \\ 4 \\ 3 \end{pmatrix} = \begin{pmatrix} -3 \\ -2 \\ 2 \end{pmatrix}$

$\overrightarrow{CB} = \boldsymbol{b} - \boldsymbol{c} = \begin{pmatrix} 6 \\ 0 \\ 3 \end{pmatrix} - \begin{pmatrix} 6 \\ 4 \\ 3 \end{pmatrix} = \begin{pmatrix} 0 \\ -4 \\ 0 \end{pmatrix}$

b) Use $\cos\theta° = \dfrac{\boldsymbol{p}\cdot\boldsymbol{q}}{|\boldsymbol{p}||\boldsymbol{q}|}$

with $\boldsymbol{p} = \begin{pmatrix} -3 \\ -2 \\ 2 \end{pmatrix}$

and $\boldsymbol{q} = \begin{pmatrix} 0 \\ -4 \\ 0 \end{pmatrix}$

$\boldsymbol{p}.\boldsymbol{q} = -3 \times 0 + (-2) \times (-4) + 2 \times 0$
$= 8$

2. b) (continued)

$$|\boldsymbol{p}| = \sqrt{(-3)^2 + (-2)^2 + 2^2}$$
$$= \sqrt{9+4+4} = \sqrt{17}$$
$$|\boldsymbol{q}| = \sqrt{0^2 + (-4)^2 + 0^2} = \sqrt{16} = 4$$

So $\cos\theta^\circ = \dfrac{8}{\sqrt{17}\times 4}$

$$\Rightarrow \theta^\circ = \cos^{-1}\left(\frac{8}{4\sqrt{17}}\right) = 60\cdot 98\ldots^\circ$$

The angle between edges CA and CB is approximately 61·0° (to 1 decimal place).

3. a) For Jan 1 2000 set $t = 0$

$$p = 6\times e^{0\cdot 0138\times 0}$$
$$= 6\times e^0 = 6\times 1 = 6$$

The population was 6 billion.

b) Solve: $12 = 6\,e^{0\cdot 0138t}$

$$\Rightarrow 2 = e^{0\cdot 0138t}$$
$$\Rightarrow \log_e 2 = 0\cdot 0138t$$
$$\Rightarrow t = \frac{\log_e 2}{0\cdot 0138} = 50\cdot 22\ldots$$

So during 2050 the population reaches 12 billion.

At the start of 2051 the population will be more than double its level in 2000.

4. a) $\cos(a+b)^\circ$

$$= \cos a^\circ \cos b^\circ - \sin a^\circ \sin b^\circ$$
$$= \frac{3}{5}\times\frac{5}{13} - \frac{4}{5}\times\frac{12}{13}$$
$$= \frac{15}{65} - \frac{48}{65} = -\frac{33}{65}$$

b) $\sin(a+b)^\circ$

$$= \sin a^\circ \cos b^\circ + \cos a^\circ \sin b^\circ$$
$$= \frac{4}{5}\times\frac{5}{13} + \frac{3}{5}\times\frac{12}{13}$$
$$= \frac{20}{65} + \frac{36}{65} = \frac{56}{65}$$

c) $\tan(a+b)^\circ = \dfrac{\sin(a+b)^\circ}{\cos(a+b)^\circ}$

$$= -\frac{56}{33}$$

5. $f(g(x)) = f(2x+1)$

$$= \frac{2}{(2x+1)+1} = \frac{2}{2x+2}$$
$$= \frac{2}{2(x+1)} = \frac{1}{x+1}$$

Also $\dfrac{1}{2}f(x) = \dfrac{1}{2}\times\dfrac{2}{x+1}$

$$= \frac{2}{2(x+1)} = \frac{1}{x+1}$$

So $f(2x+1) = \dfrac{1}{2}f(x)$

# Unit 2 sample exam question solutions

## Non-calculator

1. a) i)

$$\begin{array}{r|rrrr} -1 & 2 & 3 & 0 & -1 \\ & & -2 & -1 & 1 \\ \hline & 2 & 1 & -1 & 0 \end{array}$$

So since $h(-1) = 0$
$x + 1$ is a factor of $h(x)$

ii) $h(x) = (x+1)(2x^2 + x - 1)$
$= (x+1)(2x-1)(x+1)$
$= (x+1)^2(2x-1)$

iii) $h(x) = 0 \Rightarrow (x+1)^2(2x-1) = 0$
$\Rightarrow x+1 = 0$ or $2x - 1 = 0$
$\Rightarrow x = -1$ or $x = \frac{1}{2}$

b) The curves intersect where $f(x) = g(x)$

$\Rightarrow 2x^3 + x - 3 = -3x^2 + x - 2$
$\Rightarrow 2x^3 + x - 3 + 3x^2 - x + 2 = 0$
$\Rightarrow 2x^3 + 3x^2 - 1 = 0$
$\Rightarrow x = -1$ or $x = \frac{1}{2}$ (from (*a*))

For a common tangent $f'(x) = g'(x)$ at the point of intersection.

$f(x) = 2x^3 + x - 3$
$\Rightarrow f'(x) = 6x^2 + 1$
$g(x) = -3x^2 + x - 2$
$\Rightarrow g'(x) = -6x + 1$
$f'(-1) = 6 \times (-1)^2 + 1 = 7$
$g'(-1) = -6 \times (-1) + 1 = 7$
$f'\left(\frac{1}{2}\right) = 6 \times \left(\frac{1}{2}\right)^2 + 1 = \frac{5}{2}$
$g'\left(\frac{1}{2}\right) = -6 \times \frac{1}{2} + 1 = -2$

So only $x = -1$ gives a common tangent.

The $y$-coordinate of T is given by:
$f(-1) = 2 \times (-1)^3 + (-1) - 3$
$= -2 - 1 - 3 = -6$
So T$(-1, -6)$

2. $$\int_{-1}^{1} \frac{6x^3 - x}{3x^3}\,dx = \int_{-1}^{1} \frac{6x^3}{3x^3} - \frac{x}{3x^3}\,dx$$
$$= \int_{-1}^{1} 2 - \frac{x^{-2}}{3}\,dx$$
$$= \left[2x - \frac{x^{-1}}{3\times(-1)}\right]_{-1}^{1}$$
$$= \left[2x + \frac{1}{3x}\right]_{-1}^{1}$$
$$= \left(2\times1 + \frac{1}{3\times1}\right) - \left(2\times(-1) + \frac{1}{3\times(-1)}\right)$$
$$= 2 + \frac{1}{3} + 2 + \frac{1}{3} = 4\tfrac{2}{3}$$

3. $y = \frac{1}{\sin^2 x} = \frac{1}{(\sin x)^2} = (\sin x)^{-2}$

So $\frac{dy}{dx} = -2(\sin x)^{-3} \times \cos x = -\frac{2\cos x}{\sin^3 x}$

4. Let $y = \frac{2x+6}{\sqrt{x}} = \frac{2x+6}{x^{\frac{1}{2}}} = \frac{2x^1}{x^{\frac{1}{2}}} + \frac{6}{x^{\frac{1}{2}}}$
$= 2x^{\frac{1}{2}} + 6x^{-\frac{1}{2}}$

So $\frac{dy}{dx} = \frac{1}{2} \times 2x^{-\frac{1}{2}} - \frac{1}{2} \times 6x^{-\frac{3}{2}}$
$= x^{-\frac{1}{2}} - 3x^{-\frac{3}{2}}$

5. Differentiating a cubic expression produces a quadratic expression. So $y = f'(x)$ is a parabola. Stationary points occur when $x = 0$ and $x = b$ ($f'(0) = 0$ and $f'(b) = 0$), i.e. $y = f'(x)$ intersects the axis for these values of $x$:

   Sketch of $y = f'(x)$

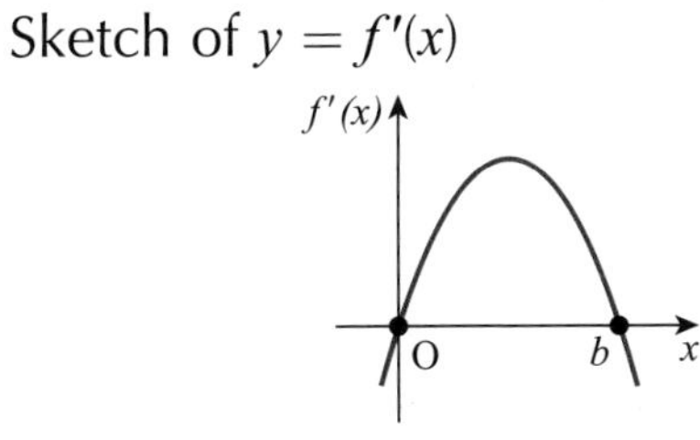


6. a) Left-hand side

   $= (\sin A + \cos B)^2 + (\cos A + \sin B)^2$

   $= \sin^2 A + 2\sin A \cos B + \cos^2 B + \cos^2 A + 2\cos A \sin B + \sin^2 B$

   $= (\sin^2 A + \cos^2 A) + (\sin^2 B + \cos^2 B) + 2(\sin A \cos B + \cos A \sin B)$

   $= 1 + 1 + 2\sin(A + B)$

   $= 2 + 2\sin(A + B)$

   $=$ Right-hand side

   b) The equation

   $(\sin A + \cos B)^2 + (\cos A + \sin B)^2 = 3$ becomes $2 + 2\sin(A + B) = 3$

   $\Rightarrow 2\sin(A + B) = 1$

   $\Rightarrow \sin(A + B) = \dfrac{1}{2}$

   [$(A + B)$ is in the 1st or 2nd quadrants and 1st quadrant angle is $\frac{\pi}{6}$]

   So $(A + B) = \dfrac{\pi}{6}$ or $\pi - \dfrac{\pi}{6} = \dfrac{5\pi}{6}$

7. $2y - 3x = 6 \Rightarrow 2y = 3x + 6 \Rightarrow y = \dfrac{3}{2}x + 3$

   The gradient of this line is $\frac{3}{2}$

   tangent

   gradient at this point is $\frac{3}{2}$

   $$y = 2\sqrt{x+1} = 2(x+1)^{\frac{1}{2}}$$

   $$\Rightarrow \frac{dy}{dx} = \frac{1}{2} \times 2(x+1)^{-\frac{1}{2}} \times 1$$

   $$= (x+1)^{-\frac{1}{2}} = \frac{1}{\sqrt{(x+1)}}$$

   This is the gradient formula for the curve.

   So for $\dfrac{dy}{dx} = \dfrac{3}{2} \Rightarrow \dfrac{1}{\sqrt{(x+1)}} = \dfrac{3}{2}$

   $\Rightarrow 2 = 3\sqrt{(x+1)}$

   $\Rightarrow 4 = 9 \times (x+1)$ (after squaring both sides)

   $\Rightarrow x + 1 = \dfrac{4}{9}$

   $\Rightarrow x = \dfrac{4}{9} - 1 = -\dfrac{5}{9}$

# Sample exam question solutions

## Calculator allowed

1. a) $k\cos(x + \alpha)°$

$= k\cos x° \cos\alpha° - k\sin x° \sin\alpha°$

compare $f(x) = 1\cos x° - 5\sin x°$

This gives:

$k\cos\alpha° = 1$
$k\sin\alpha° = 5$

Since $k > 0$ this means both $\sin\alpha°$ and $\cos\alpha°$ are positive. $\alpha°$ is in the 1st quadrant.

Use $\frac{\sin\alpha}{\cos\alpha} = \tan\alpha$

$$\frac{k\sin\alpha°}{k\cos\alpha°} = \frac{5}{1}$$

$\Rightarrow \tan\alpha° = 5$

$\Rightarrow \alpha \doteqdot 78\cdot7$

Use $\sin^2\alpha + \cos^2\alpha = 1$

$(k\sin\alpha°)^2 + (k\cos\alpha°)^2 = 5^2 + 1^2$

$k^2\sin^2\alpha° + k^2\cos^2\alpha° = 25 + 1$

$k^2(\sin^2\alpha° + \cos^2\alpha°) = 26$

$k^2 \times 1 = 26$

$k = \sqrt{26}\ (k > 0)$

So $\cos x° - 5\sin x° = \sqrt{26}\cos(x + 78\cdot7)°$

b) $f(x) = 1$ becomes

$\sqrt{26}\cos(x + 78\cdot7)° = 1$

$\Rightarrow \cos(x + 78\cdot7)° = \frac{1}{\sqrt{26}}$

$= 0\cdot1961\ldots$

The angle $(x + 78\cdot7)°$ is in the 1st or 4th quadrants.

So $x + 78\cdot7 = 78\cdot7$

or $x + 78\cdot7 = 360 - 78\cdot7$

giving $x = 0$ or $x = 202\cdot6$

c) For $x$-axis intercept set $y = 0$ i.e. $f(x) = 0$

so $\sqrt{26}\cos(x + 78\cdot7)° = 0$

giving $\cos(x + 78\cdot7)° = 0$

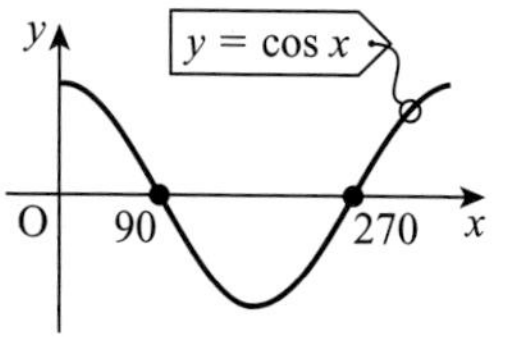

so $x + 78\cdot7 = 90$ or $x + 78\cdot7 = 270$

If $x + 78\cdot7 = 90$

$\Rightarrow x = 90 - 78\cdot7$

$= 11.3$

This is not in the required range

If $x + 78\cdot7 = 270$

$\Rightarrow x = 270 - 78\cdot7$

$= 191.3$

Thus $a = 191\cdot3$

2. a) Show that $\cos^2 x° - \cos 2x° = 1 - \cos^2 x°$

Left-hand side $= \cos^2 x° - \cos 2x°$

$= \cos^2 x° - (2\cos^2 x° - 1)$

$= \cos^2 x° - 2\cos^2 x° + 1$

$= 1 - \cos^2 x°$

$=$ Right-hand side

so $\cos^2 x°° - \cos 2x° = 1 - \cos^2 x°$

b) The equation $3\cos^2 x° - 3\cos 2x° = 8\cos x°$ becomes $3(\cos^2 x° - \cos 2x°) = 8\cos x°$

$\Rightarrow 3(1 - \cos^2 x°) = 8\cos x°$ (using part **a**)

$\Rightarrow 3 - 3\cos^2 x° = 8\cos x°$

$\Rightarrow 3\cos^2 x° + 8\cos x° - 3 = 0$

$\Rightarrow (3\cos x° - 1)(\cos x° + 3) = 0$

so $3\cos x° - 1 = 0$ or $\cos x° + 3 = 0$

$\cos x° = \frac{1}{3}$ ($x$ is in 1st or 4th quadrants)

or $\cos x° = -3$ This equation has no solutions since $\cos x°$ in never less than $-1$.

so $x = 70\cdot5$ or $360 - 70\cdot5$

$= 289\cdot5$

$289\cdot5$ is not in the required range. The only valid solution is $x = 70\cdot5$

3. $(kx + 2)(x + 3) = 8$

$kx^2 + 3kx + 2x + 6 = 8$

$kx^2 + (3k + 2)x - 2 = 0$

Discriminant $= (3k + 2)^2 - 4 \times k \times (-2)$

$= 9k^2 + 12k + 4 + 8k$

$= 9k^2 + 20k + 4$

For equal roots Discriminant $= 0$

so $9k^2 + 20k + 4 = 0$

$\Rightarrow (9k + 2)(k + 2) = 0$

$\Rightarrow 9k + 2 = 0$ or $k + 2 = 0$

$\Rightarrow k = -\frac{2}{9}$ or $k = -2$

4. a) $y = 3 + 2x^2 - x^4 \Rightarrow \frac{dy}{dx} = 4x - 4x^3$

For stationary points set $\frac{dy}{dx} = 0$

so $4x - 4x^3 = 0 \Rightarrow 4x(1 - x^2) = 0$

$\Rightarrow 4x(1 - x)(1 + x) = 0$

$\Rightarrow x = 0$ or $x = 1$ or $x = -1$

When $x = 0$, $y = 3 + 2 \times 0^2 - 0^4 = 3$ giving $(0, 3)$

When $x = 1$, $y = 3 + 2 \times 1^2 - 1^4 = 4$ giving $(1, 4)$

When $x = -1$, $y = 3 + 2 \times (-1)^2 - (-1)^4 = 4$ giving $(-1, 4)$

The stationary points are $(-1, 4)$, $(0, 3)$ and $(1, 4)$

b) $(1 + x)$

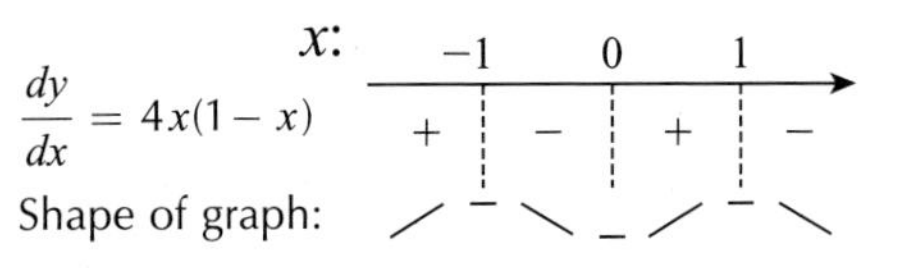


So $(-1, 4)$ and $(1, 4)$ are maximum stationary points and $(0, 3)$ is a minimum stationary point.

5. a) Let $3\cos\theta° - \sin\theta° = k\cos(\theta° + \alpha)°$: so

$3\cos\theta° - 1\sin\theta°$

$= k\cos\theta° \cos\alpha° - k\sin\theta° \sin\alpha°$

Comparing these two expressions gives:

$$\left.\begin{aligned} -k\sin\alpha° &= -1 \\ k\cos\alpha° &= 3 \end{aligned}\right\} \Rightarrow \left.\begin{aligned} k\sin\alpha° &= 1 \\ k\cos\alpha° &= 3 \end{aligned}\right\}$$

Since both $\sin\alpha°$ and $\cos\alpha°$ are positive, $\alpha°$ is in the 1st quadrant.

$\frac{k\sin\alpha°}{k\cos\alpha°} = \frac{1}{3} \Rightarrow \tan\alpha° = \frac{1}{3}$

so $\alpha° = 18{\cdot}4$ (to 3 sig. figs)

$(k\sin\alpha°)^2 + (k\cos\alpha°)^2 = 1^2 + 3^2$

$k^2\sin^2\alpha° + k^2\cos^2\alpha° = 1 + 9$

$k^2(\sin^2\alpha° + \cos^2\alpha°) = 10$

$k^2 \times 1 = 10$

$k = \sqrt{10}$ $(k > 0)$

So $3\cos\theta° - \sin\theta° = \sqrt{10}\cos(\theta + 18{\cdot}4)°$

b) $f(\theta) = \sqrt{10}\cos(\theta + 18{\cdot}4)°$

Maximum value is $\sqrt{10}$ when

$\theta + 18{\cdot}4 = 0$ or $360$

i.e. $\theta = -18{\cdot}4$ or $360 - 18{\cdot}4 = 341{\cdot}6$

but $0 \le \theta < 360$

so $\theta = 341{\cdot}6$ is the only possibility

Minimum value is $-\sqrt{10}$ when

$\theta + 18{\cdot}4 = 180$, i.e. $\theta = 180 - 18{\cdot}4$

$\Rightarrow \theta = 161{\cdot}6$

c) $\sqrt{10}(3\cos\theta° - \sin\theta°) + 10$

$= \sqrt{10} \times \sqrt{10}\cos(\theta° + 18{\cdot}4)° + 10$

So minimum value is

$\sqrt{10} \times \sqrt{10} \times (-1) + 10$

$= -10 + 10 = 0$

6. a) $f(x) = x^4 + 4x^3 + 5x^2 + 14x + 24$

| −2 | 1 | 4 | 5 | 14 | 24 |
|---|---|---|---|---|---|
| | | −2 | −4 | −2 | −24 |
| | 1 | 2 | 1 | 12 | 0 |

So $f(x) = (x + 2)(x^3 + 2x^2 + x + 12)$

| −3 | 1 | 2 | 1 | 12 |
|---|---|---|---|---|
| | | −3 | 3 | −12 |
| | 1 | −1 | 4 | 0 |

giving
$f(x) = (x + 2)(x + 3)(x^2 - x + 4)$

Thus $a = 2$ and $b = 3$
(or $a = 3$ and $b = 2$)

b) The equation $f(x) = 0$ becomes

$(x + 2)(x + 3)(x^2 - x + 4) = 0$

So $x + 2 = 0$ or $x + 3 = 0$ or
$x^2 - x + 4 = 0$

$\Rightarrow x = -2$ or $x = -3$

For $x^2 - x + 4 = 0$

the discriminant $= (-1)^2 - 4 \times 1 \times 4$

$= 1 - 16 = -15$

Since this is negative there are no real roots.

The only two real roots of $f(x) = 0$ are $x = -2$ and $x = -3$

7. $5 \times 3^\alpha = 2 \Rightarrow 3^\alpha = \dfrac{2}{5} \Rightarrow 3^\alpha = 0{\cdot}4$

So $\log_{10}3^\alpha = \log_{10}0{\cdot}4$

$\Rightarrow \alpha\log_{10}3 = \log_{10}0{\cdot}4$

$$\Rightarrow \alpha = \frac{\log_{10}0{\cdot}4}{\log_{10}3} = -0{\cdot}834\ldots$$

Thus $\cos^2 x - \sin^2 x = -0{\cdot}834\ldots$

$\Rightarrow \cos 2x = -0{\cdot}834\ldots$

($2x$ is in 2nd or 3rd quadrants)

(1st quadrant angle is $0{\cdot}584\ldots$ radians)

So $2x = \pi - 0{\cdot}584\ldots$ or $\pi + 0{\cdot}584\ldots$

So $2x = 2{\cdot}557\ldots$ only the smallest positive value is required.

So $x = 1{\cdot}278\ldots$

The required value is:

$x \doteqdot 1{\cdot}28$ (to 3 sig. figs)

# Unit 3 sample exam question solutions

## Non-calculator

1. Coordinates of M:

$$M\left(\frac{-1+3}{2}, \frac{3+(-1)}{2}\right)$$

$$= M\left(\frac{2}{2}, \frac{2}{2}\right) = M(1, 1)$$

Gradient of line:

$$m_{BC} = \frac{3-0}{-1-(-2)} = \frac{3}{-1+2} = \frac{3}{1} = 3$$

So $m_{\perp} = -\frac{1}{3}$

Equation of line:

Point on line is M(1, 1) and gradient is $-\frac{1}{3}$

Required equation is $y-1 = -\frac{1}{3}(x-1)$

So $3y - 3 = -(x - 1)$

So $3y - 3 = -x + 1$

giving $3y + x = 4$

2. a) The multiplier $a$ must lie between $-1$ and 1, i.e. $-1 < a < 1$, for a limit to exist.

b) Since $u_{n+1} = au_n + b$ and $u_1 = 5$

Then $u_2 = au_1 + b = a \times 5 + b$
$= 5a + b$

So $5a + b = 8$

also $u_3 = au_2 + b$

$= a \times 8 + b$

$= 8a + b$

So $8a + b = 9{\cdot}5$

Now solve $\left.\begin{array}{l} 5a + b = 8 \\ 8a + b = 9{\cdot}5 \end{array}\right\} \Rightarrow 3a = 1{\cdot}5$

So $a = \frac{1{\cdot}5}{3} = 0{\cdot}5$

So $5a + b = 8$

gives $5 \times 0{\cdot}5 + b = 8$

So $2{\cdot}5 + b = 8 \Rightarrow b = 5{\cdot}5$

c) The recurrence relation is
$u_{n+1} = 0{\cdot}5u_n + 5{\cdot}5$

Let the limit of the sequence be L then

$L = 0{\cdot}5L + 5{\cdot}5$

$L - 0{\cdot}5L = 5{\cdot}5 \Rightarrow 0{\cdot}5L = 5{\cdot}5$

$\Rightarrow L = \frac{5{\cdot}5}{0{\cdot}5} = 11$

3. Consider $x^2 + y^2 - 2x - 6y + 1 = 0$.

Centre is (1, 3)

$$\text{Radius} = \sqrt{1^2 + 3^2 - 1}$$

$$= \sqrt{9} = 3$$

Now use this information on the given diagram:

From the diagram...

The centre of the upper small circle is (1, 6) and its radius is 1.

So the required equation is:

$(x - 1)^2 + (y - 6)^2 = 1$

4. $$\int_{\frac{\pi}{6}}^{\frac{\pi}{2}} \sin x\, dx = [-\cos x]_{\frac{\pi}{6}}^{\frac{\pi}{2}}$$

$$= \left(-\cos\frac{\pi}{2}\right) - \left(-\cos\frac{\pi}{6}\right)$$

$$= -0 + \frac{\sqrt{3}}{2} = \frac{\sqrt{3}}{2}$$

Area of the shaded region is $\frac{\sqrt{3}}{2}$ unit²

5. a) For A(1, 2) and B(–3, –1) then

$$AB = \sqrt{(1-(-3))^2 + (2-(-1))^2}$$
$$= \sqrt{16+9} = \sqrt{25} = 5$$

For A(1, 2) and C(4, –2) then

$$AC = \sqrt{(1-4)^2 + (2-(-2))^2}$$
$$= \sqrt{9+16} = \sqrt{25} = 5$$

so AB = AC and triangle ABC is isosceles

b) $$m_{AB} = \frac{2-(-1)}{1-(-3)} = \frac{3}{4}$$

$$\tan\theta = \frac{3}{4}$$

A
θ
x
B

## Calculator allowed

1.

y
Q(−1, 4)
R(4, 5)
O
S
x
P(−2, −1)

$$m_{SR} = m_{PQ} = \frac{4-(-1)}{-1-(-2)} = \frac{5}{1} = 5$$

So gradient of line is 5 and point on line is R(4, 5)

Equation is $y - 5 = 5(x - 4)$

$y - 5 = 5x - 20$

$y = 5x - 15$

2. a) Consider $y = x^3 - 2x + 1$

$$\Rightarrow \frac{dy}{dx} = 3x^2 - 2$$

when $x = 1$ (at point P) then

$$\frac{dy}{dx} = 3 \times 1^2 - 2 = 3 - 2 = 1$$

The gradient of the tangent is 1, point on tangent is P(1, 0), so the equation is:

$y - 0 = 1(x - 1)$

$\Rightarrow y = x - 1$

b) To find Q, the point of intersection, solve:

$$\left.\begin{aligned} y &= x^3 - 2x + 1 \\ y &= x - 1 \end{aligned}\right\} \Rightarrow x^3 - 2x + 1 = x - 1$$

$$\Rightarrow x^3 - 3x + 2 = 0$$

To factorise $x^3 - 3x + 2$ try dividing it by $x - 1$:

| 1 | 1 | 0 | −3 | 2 |
|---|---|---|---|---|
| | | 1 | 1 | −2 |
| | 1 | 1 | −2 | 0 |

So $x^3 - 3x + 2 = (x - 1)(x^2 + x - 2)$

$= (x - 1)(x + 2)(x - 1)$

The equation becomes:

$(x - 1)(x + 2)(x - 1) = 0$

So $x = 1$ or $x = -2$

$x = 1$ gives the known point of intersection P(1, 0)

when $x = -2$ then $y = x - 1$

$= -2 - 1$

$= -3$

so Q(–2, –3)

c)

$y = x^3 - 2x + 1$
$x = 1$
$y = x - 1$
$x = -2$

$$\int_{-2}^{1} (x^3 - 2x + 1) - (x - 1)\,dx$$

$$= \int_{-2}^{1} x^3 - 2x + 1 - x + 1\,dx$$

$$= \int_{-2}^{1} x^3 - 3x + 2\,dx$$

$$= \left[\frac{x^4}{4} - \frac{3x^2}{2} + 2x\right]_{-2}^{1}$$

$$= \left(\frac{1^4}{4} - \frac{3\times1^2}{2} + 2\times1\right)$$

$$-\left(\frac{(-2)^4}{4} - \frac{3\times(-2)^2}{2} + 2\times(-2)\right)$$

$$= \left(\frac{1}{4} - \frac{3}{2} + 2\right) - (4 - 6 - 4)$$

$$= \frac{1}{4} - \frac{6}{4} + 2 - 4 + 6 + 4$$

$$= -\frac{5}{4} + 8 = \frac{27}{4}$$

Required area (shaded) $= \frac{27}{4}$ unit$^2$

3.

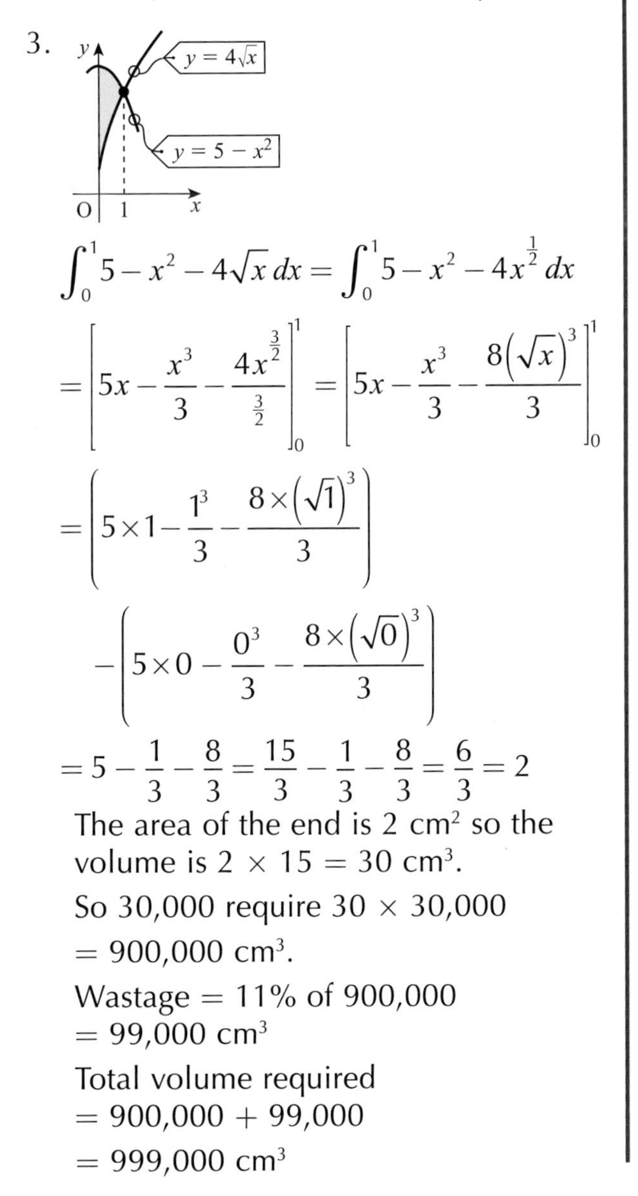


$$\int_0^1 5 - x^2 - 4\sqrt{x}\,dx = \int_0^1 5 - x^2 - 4x^{\frac{1}{2}}\,dx$$

$$= \left[5x - \frac{x^3}{3} - \frac{4x^{\frac{3}{2}}}{\frac{3}{2}}\right]_0^1 = \left[5x - \frac{x^3}{3} - \frac{8\left(\sqrt{x}\right)^3}{3}\right]_0^1$$

$$= \left(5\times1 - \frac{1^3}{3} - \frac{8\times\left(\sqrt{1}\right)^3}{3}\right)$$

$$-\left(5\times0 - \frac{0^3}{3} - \frac{8\times\left(\sqrt{0}\right)^3}{3}\right)$$

$$= 5 - \frac{1}{3} - \frac{8}{3} = \frac{15}{3} - \frac{1}{3} - \frac{8}{3} = \frac{6}{3} = 2$$

The area of the end is 2 cm$^2$ so the volume is $2 \times 15 = 30$ cm$^3$.

So 30,000 require $30 \times 30{,}000$
$= 900{,}000$ cm$^3$.

Wastage = 11% of 900,000
$= 99{,}000$ cm$^3$

Total volume required
$= 900{,}000 + 99{,}000$
$= 999{,}000$ cm$^3$

4. a) $x^2 + y^2 + 4x + 2y - 15 = 0$

Centre: $(-2, \quad -1)$

So point on AB is $(-2, -1)$ and gradient is 2 equation of AB is

$y - (-1) = 2(x - (-2))$

$\Rightarrow y + 1 = 2(x + 2)$

$\Rightarrow y = 2x + 3$

The line AB, with equation $y = 2x + 3$, has $y$-axis intercept of $(0, 3)$

when $x = 0$ and $y = 3$

$x^2 + y^2 + 4x + 2y - 15$

$= 0^2 + 3^2 + 4 \times 0 + 2 \times 3 - 15$

$= 0$

So $(0, 3)$ also lies on the circle. Hence $A(0, 3)$ is the point of intersection of the line and circle and lies on the $y$-axis.

b) now $\overrightarrow{AC} = \overrightarrow{CB}$

So $\mathbf{c} - \mathbf{a} = \mathbf{b} - \mathbf{c}$

$\mathbf{b} = 2\mathbf{c} - \mathbf{a}$

$= 2\begin{pmatrix}-2\\-1\end{pmatrix} - \begin{pmatrix}0\\3\end{pmatrix}$

$= \begin{pmatrix}-4\\-5\end{pmatrix}$

Thus $B(-4, -5)$

So

$$m_{BC} = \frac{-1-(-5)}{-2-(-4)} = \frac{4}{2} = 2 \Rightarrow m_{\perp} = -\frac{1}{2}$$

gradient of tangent is $\frac{1}{2}$, point on tangent is $B(-4, -5)$

so equation of tangent is

$y - (-5) = -\frac{1}{2}(x - (-4))$

$\Rightarrow y + 5 = -\frac{1}{2}(x + 4)$

$\Rightarrow 2y + 10 = -x - 4$

$\Rightarrow 2y + x = -14$

5. a) For the equation:

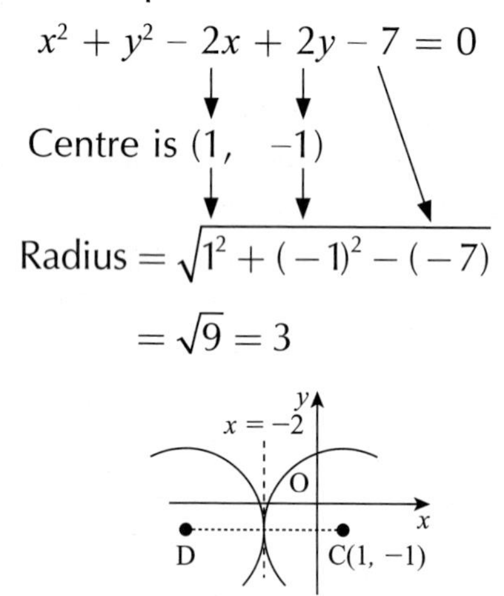


If D is the centre of the other circle then CD is parallel to the $x$-axis and CD = 6 so D(−5, −1). The radius being 3 gives equation:

$(x + 5)^2 + (y + 1)^2 = 9$

b) For $x^2 + y^2 + 4x + 2y + c = 0$

The centre is E(−2, −1)

with radius $= \sqrt{(-2)^2 + (-1)^2 - c}$

$= \sqrt{5 - c}$

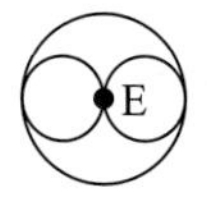


To enclose the two circles, the radius of large circle is greater than 6 units

so $\sqrt{5 - c} > 6$

$\Rightarrow 5 - c > 36$

$\Rightarrow 5 - 36 > c$

Thus $c < -31$

# Index

## Index